# Humans to Mars
## Fifty Years of Mission Planning, 1950–2000

by David S. F. Portree

NASA History Division
Office of Policy and Plans
NASA Headquarters
Washington, DC 20546

Monographs in Aerospace History Series
Number 21
February 2001

**Library of Congress Cataloging-in-Publication Data**
Portree, David S. F.
Humans to Mars: fifty years of mission planning, 1950–2000/by David S. F. Portree.
p. cm.–(Monographs in aerospace history; no. 21) (NASA publication SP)
Includes bibliographical references and index.
1. Space flight to Mars–Planning. 2. United States. National Aeronautics and Space Administration.
I. Title. II. Series. III. NASA SP ; no. 4521.

TL799.M3 P67 2000
629.45'53–dc21                                                                                                    00—062218

# Contents

# Foreword

The planet Mars has long held a special fascination and even mythic status for humans. While not the closest planet to Earth, scientists have considered it to be the planet that most closely resembles Earth and thus is the other planet in our solar system most likely to contain life. Since before the space age began, people have wondered about the "red planet" and dreamed of exploring it.

In the twentieth century, robotic spacecraft and then human space flight became a reality. Those who wanted to explore Mars in person felt that this might finally become a reality as well. The Apollo program, which put twelve Americans on the surface of the Moon, certainly encouraged the dreamers and planners who wanted to send astronauts to Mars. Indeed, many people in and out of the National Aeronautics and Space Administration (NASA) have felt that human exploration of Mars is the next logical step in human space flight after the Moon.

Clearly, however, many obstacles have remained. Human travel to and from Mars probably would take many months at best. Thus the biomedical and psychological implications of such long-duration missions are daunting. The logistics of getting enough food, water, and other supplies to Mars are also challenging at best. What would astronauts do once they got to Mars? How long would they stay on the planet's surface and how would they survive there before returning to Earth? The financial cost of sending humans to Mars would almost surely be measured in billions of dollars. Aside from technical and financial issues, there remains the political question of why we should send humans to Mars at all.

David Portree takes on these questions in this monograph. By examining the evolution of 50 mission studies over the past 50 years, he gives us a sense of the many options that Mars human space flight planners in the United States have explored. Portree covers a wide variety of ideas for human exploration of Mars, ranging from Wernher von Braun's of the 1950s to the Space Exploration Initiative of 1989. These concepts, culled from a much larger pool of studies, range from hugely ambitious flotilla-style expeditions to much leaner plans. This monograph provides historians, space policy practitioners, and other readers with a very valuable overview of how much planning has already been done. If humans do go to Mars any time in the near future, it is quite conceivable that their mission profile will resemble one of the plans described here.

A number of people helped to produce this monograph. In the NASA History Office, M. Louise Alstork edited and proofread the manuscript, while Stephen J. Garber and Nadine J. Andreassen also assisted with editing and production. The Printing and Design Office developed the layout and handled the printing. Shawn Flowers and Lisa Jirousek handled the design and editing, respectively, while Stanley Artis and Warren Owens saw this work through the publication process.

This is the twenty-first in a series of special studies prepared by the NASA History Office. The Monographs in Aerospace History series is designed to provide a wide variety of aerospace history investigations. These publications are intended to be tightly focused in terms of subject, relatively short in length, and reproduced in an inexpensive format to allow timely and broad dissemination to researchers. Thus they hopefully serve as useful starting points for others to do more in-depth research on various topics. Suggestions for additional publications in the Monographs in Aerospace History series are welcome.

Roger D. Launius
Chief Historian
National Aeronautics and
Space Administrations
October 25, 2000

# Preface

The story of the dreams and the unbuilt space-ships for flights to Mars should be recorded so that in the future people can examine past ideas of space travel just as we can examine the unconsummated ideas of Leonardo da Vinci by reading his notebooks. Years from now people should be able to decide for themselves whether they want to go to Mars or if they prefer to stay earthbound. But let us not destroy the dream, simply because we do not wish to pursue it ourselves. (Edward Ezell, 1979)[1]

In the past half century, visionary engineers have made increasingly realistic plans for launching astronauts to Mars to explore the planet. This monograph traces the evolution of these plans, taking into account such factors as on-going technological advancement and our improving knowledge of the red planet.

More than 1,000 piloted Mars mission studies were conducted inside and outside NASA between about 1950 and 2000. Many were the product of NASA and industry study teams, while others were the work of committed individuals or private organizations. Due to space limitations, only 50 mission studies (one per year, or less than 5 percent of the total) are described in this monograph. The studies included are believed to be representative of most of the technologies and techniques associated with piloted Mars exploration.[2]

In addition to tracing the evolution of mission concepts, this monograph examines piloted Mars mission planning from a policy standpoint. Mars plans are affected by their societal context and by the policies that grow from that context. When the human species eventually decides to send a piloted mission to Mars, the political environment in which it develops will have at least as much impact on its shape as technology, human factors, and the Martian and interplanetary environments. Hence, it stands to benefit the space technologist to study the ways in which policy has shaped (and thwart-

ed) past Mars plans. This idea may seem obvious to some readers, yet the history of piloted Mars mission planning shows that this truism has often been ignored or imperfectly understood, usually to the detriment of Mars exploration.

This history should be seen as a tool for building toward a future that includes piloted Mars exploration, not merely as a chronicle of events forgotten and plans unrealized. The author hopes to update and expand it in 15 or 20 years so that it tells a story culminating in the first piloted Mars mission. Perhaps a university student reading this monograph today will become a member of the first Mars mission crew tomorrow.

The author gratefully acknowledges the assistance provided by the following: Robert Ash, Donald Beattie, Ivan Bekey, John Charles, Benton Clark, Aaron Cohen, John Connolly, Mark Craig, Dwayne Day, Michael Duke, Louis Friedman, Kent Joosten, Paul Keaton, Geoffrey Landis, John Logsdon, Humboldt Mandell, Wendell Mendell, George Morgenthaler, Annie Platoff, Marvin Portree, Gordon Woodcock, and Robert Zubrin. Thanks also to the Exploration Faithful at NASA's Johnson Space Center for their insights and encouragement these past several years. Finally, thanks to Roger D. Launius, NASA Senior Historian, for soliciting this work and providing overall guidance.

David S. F. Portree
Houston, Texas, September 2000

## A Note on Measurement

In this monograph, measurements are given in the units used in the original study. Tons are U.S. tons (short tons) unless specified as metric tons. Measurements not associated with a specific study are given in the metric system.

# Chapter 1: On the Grand Scale

Will man ever go to Mars? I am sure he will—but it will be a century or more before he's ready. In that time scientists and engineers will learn more about the physical and mental rigors of interplanetary flight—and about the unknown dangers of life on another planet. Some of that information may become available within the next 25 years or so, through the erection of a space station above the Earth . . . and through the subsequent exploration of the [M]oon. (Wernher von Braun, 1954)[1]

## Von Braun in the Desert

At the beginning of serious Mars expedition planning in the United States stands German rocket pioneer Wernher von Braun. From 1945 to 1950, von Braun was interned at White Sands Proving Ground in New Mexico with about 60 other German rocket engineers spirited out of Nazi Germany by the U.S. Army at the end of the Second World War. Under Hitler, they had developed the first large liquid-propellant rocket, the V-2 missile, at the Nazi rocket base of Peenemünde; in the United States, they shared their missilery experience by preparing and launching captured V-2s under Army supervision.

In 1947 and 1948, to relieve boredom, von Braun wrote a novel about an expedition to Mars. Frederick Ordway and Mitchell Sharpe wrote in their history The Rocket Team that von Braun's novel "proved beyond doubt that its author was an imaginative scientist but an execrable manufacturer of plot and dialog."[2] Perhaps understandably, the novel never saw print. In 1952, however, its appendix, a collection of mathematical proofs supporting its spacecraft designs and mission plan, was published in West Germany as Das Marsprojekt. The University of Illinois Press published the English-language edition as The Mars Project a year later.[3] By then von Braun and many of his German colleagues were civilian employees of the Army Ballistic Missile Agency (ABMA) at Redstone Arsenal in Huntsville, Alabama.

Von Braun described a Mars expedition "on the grand scale," with ten 4,000-ton ships and 70 crewmembers.[4] He assumed no Earth-orbiting space station assembly base. His spacecraft were assembled from parts launched by three-stage winged ferry rockets. Nine hundred fifty ferry flights would be required to assemble the Mars "flotilla" in Earth orbit. Von Braun esti-

mated that each ferry rocket would need 5,583 tons of nitric acid and alcohol propellants to place about 40 tons of cargo into orbit, so a total of 5,320,000 tons of propellants would be required to launch all ten Mars ships. To provide a sense of scale he pointed out that "about 10 per cent of an equivalent quantity of high octane aviation gasoline was burned during the six months' operation of the Berlin Airlift" in 1948-49.[5] Von Braun estimated total propellant cost for launching the expedition into Earth orbit at $500 million.

Seven vessels in von Braun's plan were assemblages of girders and spheres without streamlining designed for the round-trip Mars voyage. Incapable of landing, they featured inflatable fabric propellant tanks and personnel spheres. Three one-way ships would each have a winged landing glider in place of a personnel sphere. At the appointed time, the flotilla's rocket engines would ignite to put the ships on a minimum-energy Earth-to-Mars trajectory. As Earth shrank behind, the Mars ship crews would discard empty Earth-departure propellant tanks and settle in for an eight-month weightless coast.

The members of the first Mars expedition would be the first humans to see the planet up close. No robotic explorers would precede them; von Braun did not anticipate the technological advancements that enabled automated explorers.

From Mars orbit they would turn telescopes toward Mars' equator to select a site for a surface exploration base camp. The first Mars landing site, however, would be determined at the time the expedition left Earth. One landing glider would deorbit and glide to a sliding touchdown on skids on one of the polar ice caps. Von Braun chose the polar caps because he believed them to be the only places on Mars where the crew could be certain of finding a smooth landing site. In von Braun's plan, the first people on Mars would abandon their glider on the ice cap and conduct a heroic 4,000-mile overland trek to the chosen base camp site on Mars' equator. There they would build a landing strip for the pair of wheeled gliders waiting in orbit. This Mars landing approach is unique to von Braun's work.

The wheeled gliders would touch down bearing the balance of the surface exploration party, leaving a skeleton crew in orbit to tend the seven remaining ships. As soon as the glider wheels stopped, the explorers would unbolt their delta wings and hoist their V-2-shaped fuselages upright to stand on their tail fins, ready for blast off back to the ships in Mars orbit in case of emergency. They would then set up an inflatable habitat, their base of operations for a 400-da y survey of Mars' deserts. They would gather samples of native flora and fauna and explore the mys terious linear "canals" glimpsed through Earth-based telescopes since the late nineteenth century. The journey bac k to Earth orbit would mirror the flight to Mars. Total expedition duration would be about three years.

In almost every Mars expedition plan from von Braun to the present, propellant has been potentially the single heaviest expedition element. Von Braun attempted to minimize total Mars expedition weight—and thus the number of expensive rockets required to launch the expedition from Earth's surface—by reducing as much as possible the amount of propellant needed to boost the expedition from Earth to Mars and back again. His mission profile—a minimum-energy Mars transfer, followed by a long stay during which the planets moved into position for a minimum-energy transfer back to Earth—was the chief means he used to reduce required propellant. This approach, called a conjunction-class mission, will be described in more detail in chapter 3. Extensive use of inflatable fabric structures also helped reduce spacecraft weight, though von Braun invoked them primarily because they could be folded to fit within the cargo bay of his hypothetical ferry rocket.

Use of multiple Mars ships and landers minimized risk to crew. If one ship failed, its crew could transfer to the other ships at the price of increased crowding . Von Braun's expedition plan boosted science return through large crews (including professional scientists), a small fleet of tractors for wide-ranging surface tra v-erses, and ample scientific gear . His thinking w as shaped by large Antarctic expeditions of his da y, such as Operation High J ump (1946-47), which included 4000 men, 13 ships, and 23 aircraft.[6] In the days before satellites, Antarctic explorers were largely cut off from the world, so experts and tec hnicians had to be on hand to contend with any situation that might arise . Von Braun anticipated that Mars explorers would face a similar situation.

In keeping with his assumption that little automation would be a vailable for precursor probes , von Braun's piloted ships were largely manually controlled, making large, naval-style crews mandatory. Lack of automated systems also dictated that some crewmembers remain in Mars orbit to tend the Earth-return ships .

It is interesting to compare von Braun' s vision with the Apollo lunar expeditions , when just two astronauts landed on Earth's Moon. An army of personnel, including scientists, formed part of eac h Apollo expedition, but remained behind on Earth. This separation w as made possible by communication advances von Braun did not anticipate. In 1952 von Braun stated that television transmission between Earth and a lunar expedition would be impractical.[7] Sixteen years later, Neil Armstrong's first footsteps on the dusty, cratered Sea of Tranquillity were televised live to 500 million people .

## Collier's

Von Braun's slender book of proofs was not widely distributed. His vision, however, won over the editors of the colorful Collier's weekly magazine, who commissioned him to write a series of space exploration artic les. The Collier's editor for the project, Cornelius Ryan, also solicited inputs from astronomer Fred Whipple, physicist Joseph Kaplan, physiologist Heinz Haber, United Nations lawyer Oscar Sc hachter, science writer Willy Ley, and others. Technical and astronomical art by Chesley Bonestell, Rolf Klep, and Fred Freeman brought von Braun's technical descriptions to life . Collier's, now defunct, had a circulation of three million, making it one of America's most popular magazines . Through the Collier's articles, charismatic von Braun became identified with space flight in the minds of Americans—the quintessential white-coated rocket scientist.

Collier's published eight artic les laying out a logical space program blueprint. The first, published on 22 March 1952, described von Braun's winged ferry rockets and a spinning , wheel-shaped artificial-gravity space station in Earth orbit. [8] Collier's readers reached the Moon in October 1952 [9] and explored Mars in the 30 April 1954 issue.[10] Each step in von Braun' s program built infrastructure and experience for the next.

Von Braun's Collier's Mars plan w as identical to that described in The Mars Project, except that the ten-ship Mars flotilla would be assembled near an Earth-orbiting space station. Again, von Braun assumed no robotic precursors. This time, however, telescopes located on the space station, high above Earth's obscuring atmosphere, would be used to refine knowledge of Mars and select candidate landing sites before the expedition left Earth.

Mars plans tend to focus on spacecraft, not astronauts. In the Collier's Mars article, however, von Braun explored the psychological problems of the Mars voyage. "At the end of a few months," he wrote, "someone is likely to go berserk. Little mannerisms—the way a man cracks his knuckles, blows his nose, the way he grins, talks or gestures—create tension and hatred which could lead to murder . . . [i]f somebody does crack, you can't call off the expedition and return to Earth. You'll have to take him with you." He also proposed censoring radio communication to prevent the crew from hearing dispiriting news about their hometowns.[11]

The Collier's articles were expanded into a series of four classic books. The first four chapters of the 1956 book The Exploration of Mars[12] covered the history of Mars observation and the then-current state of knowledge. Wrote von Braun and his collaborator Willy Ley: "This is the picture of Mars at mid-century: A small planet of which three-quarters is cold desert, with the rest covered with a sort of plant life that our biological knowledge cannot encompass . . ."[13] For von Braun, life on Mars was a given. In fact, von Braun's Mars was not too different from the New Mexico desert where he penned Das Marsprojekt.

Von Braun and Ley then described the Mars expedition. They conceded that it was "entirely possible . . . that within a decade or so successful tests with some sort of nuclear rocket propulsion system might be accomplished"; however, for the present, it was "exciting as well as instructive" to show that humans could reach Mars using available (1950s) technology.[14]

Their Mars expedition was a cut-price version of the 1952 Das Marsprojekt/1954 Collier's expedition, with just 12 crewmembers in two ships. Four hundred launches would put the parts, propellants, and supplies needed for the expedition into Earth orbit at the rate of two launches per day over seven months.

A single-passenger ship would complete the round-trip voyage. The craft would have an inflatable personnel sphere 26 feet across, with a control room on deck one and living quarters on decks two and three. The one-way cargo ship would carry the expedition's

single 177-ton landing glider in place of a personnel sphere. The ships together would weigh 3,740 tons before departing Earth orbit; the passenger ship would weigh only 38.4 tons when it returned to Earth orbit alone at the end of the expedition.

Upon reaching Mars, the crew would turn powerful telescopes toward proposed equatorial landing sites selected using telescopes on the space station. Equatorial sites were preferred, von Braun and Ley wrote, because they would be warmest. Citing the many kinds of surface features nearby—including two of the mysterious canals—they proposed as prime landing site candidate Margaritifer Sinus, a dark region visible in Earth-based telescopes.[15] The glider would descend to Mars with nine crewmembers on board (leaving three in orbit to mind the passenger ship's systems) and land on skids at about 120 miles per hour.

After the glider stopped, the intrepid explorers would walk out onto the wing, leap 18 feet to the ground (the equivalent of a six-foot drop in Earth gravity), and immediately prepare the ship for emergency liftoff—this despite having just spent eight months in weightlessness. They would remove the wings and use the expedition's two caterpillar tractors to hoist the bullet-shaped fuselage upright. They would then inflate a 20-foot hemispherical pressurized "tent" to serve as expedition headquarters.

After a year of Mars surface exploration, they would lift off, rejoin their compatriots in orbit, and blast for Earth. The last drops of propellants would place the ship in a 56,000-mile-high Earth orbit. A relief ship would ascend from the space station to collect the crew; they would abandon the Mars ship as a monument to the early days of planetary exploration.

## Mars Beckons

Every 26 months, the orbits of Earth and Mars bring the two planets relatively close together. At such times Mars becomes a bright red-orange "star" in Earth's skies. Because Mars appears opposite the Sun in the sky when it is closest to Earth, astronomers call such events oppositions.

# Chapter 1: On the Grand Scale

Mars has an elliptical orbit, so its distance from Earth at opposition varies. The best oppositions, when Earth is closest to Mars while the planet is closest to the Sun, occur roughly every 15 years . During the best oppositions, Mars' disk appears about twice as large in telescopes as during the poorest oppositions, when Mars is farthest from the Sun.

In the 1950s, all knowledge of Mars' conditions came from telescopic observations made during the best oppositions. Since the invention of the telescope in the early seventeenth century, astronomers eagerly awaited the best oppositions to attempt to pry new secrets from Mars. For example, the canals were first seen during the excellent 1877 opposition. When the first printing of T he Exploration of Mar s arrived on bookstore shelves, astronomers were eagerly a waiting the c lose opposition of September 1956.

The year 1956 marked the last "best" Mars opposition upon which astronomers would entirely depend for data, because humanity's relationship with Mars w as about to change. A year after that Mars opposition, on 4 October 1957, the Soviet Union launc hed Sputnik 1 into Earth orbit. Von Braun's U.S. Army rocket team launched the American response, Explorer 1, on 31 January 1958. By 1971, when again Mars shone as brightly in Earth's skies as in 1956, spacecraft reconnaissance had revolutionized how we learn about the solar system. As we will see in the coming c hapters, the 1956 and 1971 "best" oppositions neatly bracketed the early heyday of NASA Mars exploration planning.

# Chapter 2: Earliest NASA Concepts

Now it is time to take longer strides—time for a great new American enterprise—time for this nation to take a c learly leading role in space achievement . . . I believe that this nation should commit itself to ac hieving the goal, before this decade is out, of landing a man on the [M]oon and returning him safely to Earth. No single space project in this period will be more impressive to mankind, or more important for the long-range exploration of space . . . it will not be one man going to the [M]oon—it will be an entire nation. For all of us must work to put him there. (John F. Kennedy, 1961)[1]

## NASA's First Mars Study

As early as November 1957—a month after Sputnik 1 became Earth's first artificial moon—about 20 researchers at Lewis Researc h Center, a National Advisory Committee on Aeronautics (NACA) laboratory in Cleveland, Ohio, commenced research into nuclear-thermal and electric roc ket propulsion for interplanetary flight.[2] (Lewis was renamed Glenn Research Center at Lewis Field in 1999.) Such advanced propulsion systems required less propellant than chemical rockets, thus promising dramatic spacecraft weight savings. This meant fewer costly launches from Earth's surface and less Earth-orbital assembly.

Soon after the Lewis researc hers began their work, Congress and the Eisenhower administration began to work toward the creation of a U .S. national space agency in response to Soviet space c hallenges. President Dwight Eisenhower wanted a civilian agency to ensure that headline-grabbing space shots would not interfere with the serious business of testing missiles and launching reconnaissance satellites . Senator Clinton Anderson (Democrat-New Mexico) led a faction that wanted the Atomic Energy Commission (AEC) to run the space program, citing as justification its nuclear-thermal rocket experiments. Others supported expansion of NACA, the federal aeronautics researc h organization founded in 1915. On 29 July 1958, Eisenhower signed into la w legislation creating the National Aeronautics and Space Administration (NASA) from NAC A and various Department of Defense space organizations.[3]

When NASA opened its doors on 1 October 1958, Lewis became a NASA Center. The Lewis researchers sought to justify and expand their advanced propulsion work. In April 1959—two years before any human ventured into Earth orbit—they testified to Congress about their work and solicited funding for a Mars expedition study in Fiscal Year (FY) 1960. Congress granted the request, making the Lewis study the first Mars expedition study conducted under NASA auspices.[4]

The Lewis researc hers sought to develop weight estimates for Mars ships using their advanced propulsion systems. For their nuclear-thermal rocket analysis, the Lewis researchers assumed a Mars mission profile that would, by the end of 1960s , come to be virtually the standard NASA model:

> The mission begins with the vehicle system in an orbit about the Earth. Depending on the weight required for the mission, it can be inferred that the system has been delivered as a unit to orbit—or that it has been assembled in the orbit from its major constituents . . . the vehicle containing a crew of seven men is accelerated by a high-thrust nuc lear rocket engine onto the transfer trajectory to Mars . Upon arrival at Mars, the vehicle is decelerated to establish an orbit about the planet . . . a Mars Landing Vehicle containing two men descends to the Martian surface . . . . After a period of exploration these men take off from Mars using chemical-rocket power and effect a rendezvous with the orbit party. The . . . vehicle then accelerates onto the return trajectory; and, upon reaching Earth, an Earth Landing Vehicle separates and . . . decelerates to return the entire crew to the surface.[5]

For analysis purposes, the Lewis researchers targeted the 1971 launc h opportunity, when Mars' c lose proximity to Earth minimized the amount of energy (and thus propellant) needed to reac h it. They cautioned, however, that "[t]his is not meant to imply that actual trips are contemplated for this period."[6] They opted for a 420-day round trip with a 40-da y stay at Mars, and found that the optimum launch date was 19 May 1971.

As might be expected, fast Mars trips generally require more propellant (typically liquid hydrogen in the case of a nuc lear rocket) than slow trips . The more propellant required, the greater the spacecraft's weight at Earth-orbit departure . Thus, longer missions appear preferable if weight minimization is the dominant consideration in a Mars mission plan. The Lewis team noted, however, that crew risk factors had to be considered in calculating spacecraft weight. These were hard to judge because much about conditions in interplanetary space and on Mars remained unknown. In particular, they cautioned that "[c]urrent knowledge of radiation hazards is still not completely satisfactory."[7]

Explorer 1 and Explorer 3 (launched 26 March 1958) detected the Van Allen Radiation Belts surrounding Earth. Their discovery was the first glimpse of an unsuspected reef, rock, or shoal menacing navigators in the new ocean of space. It raised the profile of ionizing radiation as a possible threat to space travelers.

No longer were the thin-skinned personnel spheres of von Braun's 1950s Mars ships judged adequate. Von Braun had made little provision for limiting crew radiation exposure, though he had expressed the hope that "by the time an expedition from earth is ready to take off for Mars, perhaps in the mid-2000s . . . researchers will have perfected a drug which will enable men to endure radiation for comparatively long periods."[8]

The Lewis team did not place its trust in pharmacology. For their study, they assumed the following ionizing radiation sources: the Van Allen belts at Earth and Mars (in reality, Mars lacks radiation belts), continuous cosmic ray bombardment, solar flares, and, of course, the ship's nuclear-thermal rocket engine. Their spacecraft crew compartment, an unshielded two-deck cylinder providing 50 square feet of floor space per crewmember ("between that provided for chief petty officers and commissioned officers on submarines"), contained a heavily shielded cylindrical "vault" at its center, into which the crew would retreat during passage through the Van Allen belts, nuclear rocket operation, and large solar flares.[9] Crewmembers would also sleep in the vault; this would reduce their cosmic ray exposure during approximately one-third of each day.

Not surprisingly, the weight of radiation shielding required depended on how much radiation exposure for the crew was allowed. If major solar flares could be avoided during the 420-day voyage and a total radiation dose of 100 Roentgen Equivalent Man (REM) were permissible, then 23.5 tons of shielding would suffice,

the Lewis researchers found. If, however, one major flare could not be avoided, shielding weight jumped to 82 tons to keep the total dose below 100 REM. If only 50 REM were considered permissible and one major flare could not be avoided, shielding weight would become "enormous"—140 tons.[10] "These data," they wrote, served "to underscore . . . the importance of determining more precisely the nature and virulence of the radiation in space."[11]

The Lewis researchers determined that "short trips are as, or more, economical, in terms of weight, than long-duration missions," even though they generally required more propellant, because long trips required more heavy shielding to keep the crew within the radiation dose limit.[12] They estimated that a nuclear-thermal spaceship for a 420-day round trip in 1971 with a maximum allowable total radiation dose of 100 REM would weigh 675 tons at Earth-orbit launch.

## Twirling Ion Ships to Mars

Just as the creation of NASA was prompted by the Cold War clash between the United States and the Soviet Union, so was the goal that dominated NASA's first decade. On 12 April 1961, Soviet cosmonaut Yuri Gagarin became the first person to orbit Earth. His Vostok 1 spacecraft completed one circuit of the planet in about 90 minutes. Gagarin's flight was a blow to the new administration of President John F. Kennedy, who had narrowly defeated Eisenhower's Vice President, Richard M. Nixon, in the November 1960 elections. Gagarin's flight coincided with the embarrassing failure of a Central Intelligence Agency (CIA)-sponsored invasion of Cuba at the Bay of Pigs (17-19 April 1961).[13]

The tide of Kennedy's political fortunes began to turn on 5 May 1961, when astronaut Alan Shepard rode the Freedom 7 Mercury capsule on a suborbital hop into the Atlantic Ocean. On 25 May 1961, Kennedy capitalized on this success to seize back the political high ground. Before a special Joint Session of Congress, he called for an American to land on Earth's Moon by the end of the 1960s.

NASA had unveiled a 10-year plan in February 1960 that called for a space station and circumlunar flight before 1970, and a lunar landing a few years later. The

Agency believed that this constituted a logical program of experience-building steps.[14] Mars planners were torn over Kennedy's new timetable. On the one hand, it put Mars work on the back burner by making the Moon NASA's primary, overriding goal. On the other hand, it promised to make launch vehicles and experience needed for Mars available all the sooner.[15]

Two contenders led the pack of Apollo lunar mission modes in mid-1961—Earth-Orbit Rendezvous (EOR) and Direct Ascent. Both stood to benefit piloted Mars missions. In EOR, two or three boosters launched Moon ship modules into Earth orbit. The modules docked; then the resultant ship flew to the Moon and landed. Mars planners knew that experience gained through Moon ship assembly could be applied to Mars ship assembly. In Direct Ascent, the spacecraft flew directly from Earth's surface to the lunar surface and back. This called for an enormous launch vehicle which could be used to reduce the number of launches needed to put Mars ship parts and propellants into orbit.

NASA's Marshall Space Flight Center in Huntsville, Alabama, was responsible for developing the rockets required for lunar flight. Marshall began as the ABMA's Guided Missile Development Division. In the 1950s, the von Braun rocket team had developed some of the first U.S. missiles, including the intermediate-range Redstone, the "Americanized" version of the V-2. A Redstone variant called Jupiter-C launched the Explorer 1 satellite.

Just as Saturn was next after Jupiter among the planets, the Saturn series of rockets was next after Jupiter-C. Saturn I and Saturn IB used a cluster of Redstone/Jupiter tanks in their first stages. The engineers in Huntsville envisioned yet larger rockets. NASA's 1960 master plan called for development of an enormous "post-Saturn" rocket called Nova. Either Saturn or Nova could be used to carry out an EOR Moon mission; Nova was required for Direct Ascent.

Marshall might have performed the first NASA Mars study, but when the Lewis advanced propulsion engineers testified to Congress in 1959, the Huntsville organization was still not a part of NASA. Ernst Stuhlinger's group within the ABMA Guided Missile Development Division had commenced work on electric propulsion in 1953 and considered Mars expeditions in its design process.

In electric propulsion, a thruster applies electricity to propellant (for example, cesium), converting its atoms into positive ions. That is, it knocks an electron off each cesium atom, giving it an electric charge. The thruster then electrostatically "grips" the cesium ions and "throws" them at high speed. Electric propulsion provides constant low-thrust acceleration while expending much less propellant than chemical or nuclear-thermal propulsion, consequently reducing spacecraft weight. Low thrust, however, means low acceleration.

Stuhlinger presented a paper in Austria in 1954 describing a solar-powered electric-propulsion spacecraft with dish-shaped solar concentrators.[16] Walt Disney had contacted von Braun after reading the Collier's articles; this contact led to three space flight television programs from 1955 to 1957. Disney's Mars and Beyond, which premiered on 4 December 1957, featured Stuhlinger's distinctive umbrella-shaped nuclear-electric Mars ships, not von Braun's sphere-and-girder chemical ships.[17]

The U.S. Army, eager to retain its foothold in missilery, was loath to release the von Braun team to NASA as required by President Eisenhower. Army resistance prevented von Braun, Stuhlinger, and their colleagues from officially joining the new space agency until 1 July 1960. However, they had by then worked directly with NASA for some time—hence their input to NASA's February 1960 master plan.[18] Wernher von Braun became Marshall's first director, and Ernst Stuhlinger became director of Marshall's Research Projects Division.

Stuhlinger's 1962 piloted Mars mission design, targeted for launch in the early 1980s, would include five 150-meter long Mars ships of two types—"A" and "B"—each carrying three astronauts.[19] As in von Braun's The Mars Project, risk to crew was minimized through redundancy. The expedition could continue if as many as two ships were lost, provided they were not of the same type. One ship could return the entire 15-person expedition to Earth under crowded conditions.

The three "A" ships would carry one 70-ton Mars lander each. At Mars, an unpiloted cargo lander would detach; if it landed successfully, the explorers would land in the second lander. If the cargo lander failed, the second lander would become an unpiloted cargo lander, and the third lander would deliver the surface team. The lander crew would stay on Mars for 29 days. If the crew lander ascent stage failed to fire, the explorers could return to Mars orbit in the cargo lander ascent stage.

Stuhlinger's ships would each include a nuclear reactor producing 115 megawatts of heat. The reactor would heat a working fluid which would drive a turbine; the turbine in turn would drive a generator to supply 40 megawatts of electricity to two electric-propulsion thrusters. To reject the heat it retained after leaving the turbine, the working fluid would circulate through radiator panels with a total area of 4,300 square meters before returning to the reactor. The ship would move through space with its radiator panels edge-on to the Sun. Radiator tubes would be designed to be individually closed off to prevent a meteoroid puncture from releasing all of the ship's working fluid into space.

Each flat, diamond-shaped ship would weigh 360 tons when it switched on its electric thrusters in Earth orbit at the start of the Mars voyage—a little more than half as much as the NASA Lewis nuclear-thermal Mars ship. Of this, 190 tons (for the "B" ships) or 120 tons (for the "A" ships) would be cesium propellant. As already indicated, the price of low spacecraft weight was low acceleration—Stuhlinger's fleet would need 56 days to spiral up and out of Earth orbit; then, after a 146-day Earth-Mars transfer, it would require 21 days to spiral down to low-Mars orbit.

Stuhlinger's ships would rotate 1.3 times per minute to produce acceleration equal to one-tenth of Earth's gravity in the crew cabin. The reactor, located at the opposite end of the ship from the crew cabin, would act as an artificial gravity counterweight. Thus, the separation needed to keep the crew away from the reactor would also serve to increase spin radius.

Engineers designing artificial gravity systems must endeavor to make the spin radius as long as possible. This is because an artificial gravity system with a short spin radius must rotate more rapidly than one with a long spin radius to generate the same level of acceleration, which the crew feels as gravity. A short-radius, fast-spinning rotating system produces pronounced coriolis effects. For example, water leaving a faucet curves noticeably. Similarly, a person moving toward or away from the center of such a rotating system tends to veer sideways. Turning the head tends to produce nausea. In addition, a troublesome gravity gradient occurs

vertically along the body—the head experiences less acceleration than the feet.

Stuhlinger's electric thrusters would be mounted at the ship's center of rotation on stalks. These would rotate against the ship's spin to remain pointed in the required direction. In addition to aiding the crew, Stuhlinger noted, artificial gravity would prevent gas pockets from forming in the working fluid.[20]

Stuhlinger's design included a 50-ton, graphite-clad radiation shelter (about 15 percent of the entire weight of the ship) in the ship's crew compartment. Drinking water, propellant, oxygen cylinders, and equipment would be arranged around the shelter to provide additional shielding. The 2.8-meter-diameter, 1.9-meter-high shelter would hold a three-person ship's complement comfortably and would protect the entire 15-person expedition complement in an emergency. The crew would live in the shelter for 20 days during the outbound Van Allen belt crossing.

## The Moon Intervenes

Stuhlinger wrote that it "is generally accepted that a manned expedition to . . . Mars will be carried out soon after such an ambitious project becomes technically feasible . . . [it is] the natural follow-on project to be undertaken after the lunar program."[21] Mars planners took Kennedy at his word when he said that reaching the Moon was "important for the long-range exploration of space."

On 11 July 1962, however, NASA announced that it had selected Lunar Orbit Rendezvous (LOR) over EOR and Direct Ascent as the Apollo mission mode. Attention had turned from EOR and Direct Ascent to LOR early in 1962. LOR, a concept zealously promoted by NASA Langley Research Center engineer John Houbolt, promised the lowest lunar spacecraft weight. This enabled a lunar expedition with only a single Saturn rocket launch, making LOR the fastest, cheapest way of meeting Kennedy's end-of-decade deadline.[22]

In LOR, the lunar spacecraft—which consists of a small lander and a command ship—blasts off directly from Earth with no Earth-orbital assembly. The lander lands on the Moon, leaving the large command ship in lunar orbit. Surface exploration completed, the lander blasts off from the Moon and returns to the orbiting command ship. Spacecraft weight is reduced because only the small, light-weight lander must burn propellant to land and lift off.

It should be noted that the NASA Lewis and Stuhlinger Mars plans used the same general approach for the same reason. Landing the entire massive ship on Mars and launching it back to Earth would require impossible amounts of propellant or an impossibly small interplanetary vehicle. The standard NASA Mars plan can thus be dubbed Mars Orbit Rendezvous (MOR).

The LOR decision impacted post-Apollo ambitions. The reduction in lunar expedition mass promised by LOR removed the need for a post-Saturn Nova rocket, as well as the need to learn how to assemble large modular vehicles in Earth orbit. It thus reduced Apollo's utility as a technological stepping stone to Mars. The need to create a new justification for big rockets influenced Marshall's decision to start a new Mars study in early summer 1962. As will be seen in the next chapter, this study, known as EMPIRE, kicked off the most intense period of piloted Mars mission planning in NASA's history.

# Chapter 3: EMPIRE and After

Manned exploration of Mars is the key mission in interplanetary space flight. Man must play a key role in the exploration of Mars because the planet is relatively complex, remote, and less amenable to exploration by unmanned probes than is the [M]oon... serious interest in the Manned Mars Mission is springing up ... with many planning studies being performed by several study teams within [NASA] and within industry . . . . Perhaps the most important result emerging from the present stu dies is the indication that the Manned Mars Mission can be performed in the relatively near future with equipment and tec hniques that will for the most part be brought into operation by the Apollo Project ... the Manned Mars Mission is rapidly taking shape as the direct follo w-on to the Apollo Project. (Robert Sohn, 1964)[1]

## EMPIRE

Ernst Stuhlinger's Research Projects Division w as the smaller of two advanced planning groups in ABMA. The larger, under Heinz K oelle, became the Marshall Space Flight Center' s Future Projects Office . Until 1962, Koelle's group focused primarily on lunar programs—Koelle was, for example, principal author of the U.S. Army's 1959 Project Horizon study, which planned a lunar fort by 1967. Koelle's deputy, Harry Ruppe, also supervised a limited number of Mars studies. Ruppe had come from Germany to join the von Braun team in Huntsville in 1957.

In the 1962-1963 period, however, the Future Projects Office spearheaded NASA's Mars planning efforts . As discussed in the last chapter, Marshall's primary focus was on launc h vehicles. Advanced planning became important at Marshall in part because of the long lead times associated w ith developing new roc kets. Marshall director von Braun foresaw a time in the mid-1960s when his center might become idle if no goals requiring large boosters were defined for the 1970s. As T. A. Heppenheimer wrote in his 1999 book The Space Shuttle Decision,

> The development of the Saturn V set the pace for the entire Apollo program. This Moon rocket, however, would have to reac h an advanced state of reliability before it could be

used to carry astronauts . The Marshall staff also was responsible for development of the smaller Saturn IB that could put a piloted Apollo spacecraft through its paces in Earth orbit. Because both roc kets would ha ve to largely complete their development before Apollo could hit its stride , von Braun knew that his [C]enter would pass its peak of activity and would shrink in size at a relatively early date. He would face large la yoffs even while other NASA [C]enters would still be actively preparing for the first mission to the Moon.[2]

Mars was an obvious target for Marshall' s advanced planning. Von Braun w as predisposed tow ard Mars exploration, and landing astronauts on Mars provided ample scope for his Center to build new large boosters . The timing, however, was not good. The Moon would, if all went well, be reached by 1970—but NASA would certainly not be ready to land astronauts on Mars so soon. For one thing, planners needed more data on the Martian environment before they could design landers, space suits, and other surface systems. What Marshall needed was some kind of short-term interim program that answered questions about Mars while still providing scope for new roc ket development.

A 1956 paper by Italian astronomer Gaetano Crocco , presented at the Seventh International Astronautical Federation Congress in Rome, offered a possible way out of Marshall's dilemma.[3] Crocco demonstrated that a spacecraft could, in theory, fly from Earth to Mars, perform a reconnaissance Mars flyby , and return to Earth. The spacecraft would fire its roc ket only to leave Earth—it would coast for the remainder of the flight. The Mars flyby mission would require less than half as much energy—hence propellant—as a minimum-energy Mars stopover (orbital or landing) expedition. This meant a correspondingly reduced spacecraft weight. Total trip time for a Crocco-type Mars flyby was about one year; for the type of mission von Braun employed in The Mars Project (1953), trip time was about three years.

Flyby astronauts would be like tourists on a tour bus , seeing the sights from a distance in passing but not getting off. Crocco wrote that they would use "a telescope of moderate aperture . . . to reveal and distinguish natural [features] of the planet . . . ." He found, however, that Mars' gra vity would deflect the flyby spacecraft' s course so it missed Earth on the return leg if it flew closer to Mars than about 800,000 miles Such a distant flyby would, of course, "frustrate the exploration scope of the trip."

To permit a close flyby without using propellant, Crocco proposed that the c lose Mars flyby be followed by a Venus flyby to bend the craft' s course tow ard Earth. The Venus flyby would be an exploration bonus, Crocco wrote, allowing the crew to glimpse "the riddle which is concealed by her thic k atmosphere." Crocco calculated that an opportunity to begin an Earth-Mars-V enus-Earth flight would occur in J une 1971.[4]

From a vantage point at the start of the twenty-first century, a piloted planetary flyby seems a strange notion, yet in the 1960s NASA ga ve nearly as muc h attention to piloted Mars flybys as it did to piloted Mars landings. Piloted Mars flybys are now viewed from the perspective of more than three decades of successful automated flyby missions (as well as orbiters and landers). Of the nine planets in the solar system, only Pluto has not been subjected to flyby examination by machines. Robots can do flybys , so why entail the expense and risk to crew of piloted flybys?

Indeed, there were critics at the time the Future Projects Office launched its Early Manned Planetary-Interplanetary Roundtrip Expeditions (EMPIRE) piloted flyby/orbiter study . For example, Maxime Faget, principal designer of the Mercury capsule , coauthored an article in February 1963 whic h pointed out that a piloted Mars flyby would "demand the least [propulsive] energy . . . but will also ha ve the least scientific value" because of the short period spent near Mars. He added that data on Mars gathered through a piloted flyby would be "in many w ays no better than those which might be obtained with a properly operating , rather sophisticated unmanned probe."[5]

The key phrase in Faget's criticism is, of course, "properly operating." When the Future Projects Office launched EMPIRE in May-June 1962, robot probes did not yet possess a respectable performance record. The Mariner 2 probe carried out the first successful flyby exploration of another planet (V enus) in December 1962, midway through the EMPIRE study , but the other major U.S. automated effort, the Ranger lunar program, was off to a shaky start. That series did not enjoy its first success until Ranger 7 in J uly 1964. The first successful Mars flyby did not occur until a year

after that. In fact, one of the early justifications for piloted flybys was that the astronauts could act as care-takers for a cargo of automated probes to keep them healthy until just before they had to be released at the target planet.

Faget also believed that the "overall planning of a total spaceflight program should be based on a logical series of steps." Mercury and Gemini would provide basic experience in living and working in space, paving the way for Apollo, which would, Faget explained, "have the first real mission." After that, NASA should build an Earth-orbiting space station and possibly a lunar base .[6]

For Faget, a piloted Mars flyby mission in the 1970s was a deviation from the model von Braun popularized in the 1950s, which placed the first Mars expedition a century or more in the future . Faget avoided mentioning, however, that he had already been compelled to rationalize Kennedy's politically motivated drive for the Moon. Going by von Braun's logical blueprint, piloted lunar flight should have been postponed until after the Earth-orbiting space station w as in place.

For the EMPIRE study, three contractors studied piloted flyby and "capture" (orbiter) expeditions to Mars and Venus. Aeronutronic studied flybys[7]; Lockheed looked at flybys and, briefly, orbiters[8]; and General Dynamics focused on orbiter missions.[9] Aeronutronic's study summed up EMPIRE's three goals:

- Establish a requirement for the Nova roc ket development program.

- Provide inputs to the joint AEC-NASA nuclear rocket program, which had been established in 1960 and included a flight test program over which Marshall had technical direction.

- Explore advanced operational concepts necessary for flyby and orbiter missions.[10]

The first two goals were contradictory as far as spacecraft weight minimization w as concerned. Seeking justification for a new large roc ket provided little incentive for weight minimization, while one of the great attractions of nuclear-thermal rockets was their increased efficiency over chemical rockets, which helped minimize weight. The contractors' tendency not to tightly control spacecraft weight assisted them with crew risk minimization. For example, all three contractors sa w fit to inc lude in their

EMPIRE designs heavy spacecraft structures for generating artificial gravity.

Lockheed identified two main Mars flyby trajectory classes, which it nicknamed "hot" and "cool." In the former, the piloted flyby spacecraft would drop inside Earth's orbit (in some launch windows Venus flyby occurred), reach its farthest point from the Sun (aphelion) as it flew by Mars, and return to Earth about 18 months after launch. In the latter, the flyby spacecraft would fly out from Earth's orbit, pass Mars about 3 months after launch, reach aphelion in the Asteroid Belt beyond Mars, and return to Earth about 22 months after launch.

The Aeronutronic team opted for a "hot" trajectory. They assumed a Nova rocket capable of lifting 250 tons to Earth orbit. For comparison, the largest planned Saturn rocket, the Saturn C-5 (as the Saturn V was known at this time) was expected to launch around 100 tons. One Nova rocket would thus be able to launch the entire 187.5-ton Aeronutronic flyby spacecraft into Earth orbit.

Aeronutronic's "design point mission" had the flyby spacecraft leaving Earth orbit between 19 July 1970 and 16 August 1970, using a two-stage nuclear-thermal propulsion system. Aeronutronic's design retained the empty second-stage hydrogen propellant tanks to help shield the command center in the ship's core against radiation and meteoroids. Two cylindrical crew compartments would deploy from the core on booms; then the ship would rotate to provide artificial gravity. An AEC-developed radioisotope power source would deploy on a boom behind the ship. At the end of the flight the crew would board a lifting body Earth-return vehicle and separate from the ship. A two-stage retrorocket package would slow the lifting body to a safe Earth atmosphere reentry speed while the abandoned flyby ship sailed by Earth into orbit around the Sun.

Lockheed also emphasized a rotating design for its EMPIRE spacecraft. In the company's report, the flyby crew rode into orbit on a Saturn C-5 in an Apollo Command and Service Module (CSM) perched atop a folded, lightweight flyby spacecraft. A nuclear upper stage would put the CSM and flyby ship on course for Mars. The CSM would then separate and the flyby spacecraft would automatically unfold two long booms from either side of a hub. The CSM would dock at the end of one boom to act as counterweight for a cylindrical habitation module at the end of the other boom. When the ship rotated, the CSM and habitation module would experience acceleration the crew would feel as gravity.

The weightless hub at the center of rotation would contain chemical rockets for course correction propulsion, a radiation shelter, automated probes, and a dish-shaped solar power system. At Mars, the crew would stop the spacecraft's rotation and release the probes. At journey's end, the crew would separate from the flyby craft in the CSM, fire its rocket engine to slow down, discard its cylindrical Service Module (SM), and re-enter Earth's atmosphere in the conical Command Module (CM). The abandoned flyby craft would fly past Earth into solar orbit. Lockheed's report mentioned briefly how a Mars orbiter mission might investigate the Martian moons Phobos and Deimos.[11]

The General Dynamics report was by far the most voluminous and detailed of the three EMPIRE entries, reflecting a real passion for Mars exploration on the part of Krafft Ehricke, its principal author. Ehricke commanded tanks in Hitler's attack on Moscow before joining von Braun's rocket team at Peenemünde. He came to the U.S. in 1945 with the rest of the von Braun team but left in 1953 to take a job at General Dynamics in San Diego, California. There he was instrumental in Atlas missile and Centaur upper-stage development. In the late 1950s he became involved in General Dynamics advanced planning.

Ehricke's team looked at piloted Mars orbiter missions. These would permit long-term study of the planet from close at hand, thus answering critics who complained that piloted flybys would spend too little time near Mars. General Dynamics' 450-day Mars orbiter mission was set to launch in March 1975.

Modularized Mars ships would travel in "convoys" made up of at least one crew ship and two automated service ships. Ship systems would be "standardized as much as practical" so that the crew ship could cannibalize the service ships for replacement parts. If a meteoroid perforated a propellant tank, for example, the crew would be able to replace it with an identical tank from a service ship. The ships would carry small "tugboat" spacecraft for moving propellant tanks and other bulky spares.[12] This approach—providing many spares—helped minimize risk to crew, but would dramatically boost overall expedition weight.

General Dynamics described many possible ship configurations; what follows w as typical. The company allotted a nuc lear propulsion stage for eac h major maneuver. After performing its assigned maneuver, the stage would be cast off. Ehricke's team estimated that nuclear engine flight testing would ha ve to occur between May 1968 and April 1970 to support a Marc h 1975 expedition. The M-1 engine system would perform Maneuver-1 of the Mars expedition, escape from Earth orbit (hence its designation). The M-2 engine system would slow the ship so Mars' gra vity could capture it into Mars orbit, and M-3 would launc h the spacecraft out of Mars orbit toward Earth. The M-4 engine system would slow the ship at Earth at the expedition' s end.

Attached to the front of the M-4 stage would be the 10-foot-diameter, 75-foot-long spine module , or "neck," which served two functions: in addition to separating the astronauts from the nuc lear engines to minimize crew radiation exposure , it would place distance between the crew and the ship's center of gravity, making the artificial gravity spin radius longer.

General Dynamics opted arbitrarily for providing artificial gravity equal to 25 percent of Earth' s surface gravity and estimated that five rotations per minute was the upper limit for crew comfort. As engine systems were cast off, however, the ship's center of rotation would shift forw ard. For example, before the M-1 maneuver it would be at the aft end of the M-2 engine system, 420 feet from the ship's nose, and at the start of the M -2 maneuver it would be at the front of the M-2 system, 265 feet from the nose . As the ship grew prog ressively shorter, the spin radius would decrease, forcing faster rotation to maintain the same artificial gravity level. The report proposed joining the aft end of the crew vehicle to the end of a service vehicle during return to Earth, after the M-3 engine system was cast off, in order to place the center of rotation at the joint between the two vehic les and permit an acceptable rotation rate .

The General Dynamics crew ship design inc luded the Life Support Section (LSS) for the eight-person crew . The LSS, which would be tested attached to an Earth-orbital space station beginning in November 1968, again illustrated the intense modularity of the General Dynamics design. The 10-foot-diameter central section would be attached to the front of the spine module and would house the repair shop, food storage, and radiation-shielded Command Module (not to be confused with the Apollo CM). The Command Module would serve double duty as the ship's radiation shelter and "last redoubt" if all other habitable modules were destroyed. Crewmembers would sleep in the Command Module's lower level to reduce their overall radiation exposure. The top level would serve as the crew ship's bridge and the "blockhouse" from which the service vehicles would be remote-controlled.

Two-level, 10-foot-diameter Mission Modules would cluster around the central section to provide additional living space. Individual levels could be sealed off if penetrated by meteoroids , and entire Mission Modules could be cast off if the crew had to reduce spacecraft mass to permit return to Earth—for example, if a large amount of propellant were lost and could not be replaced from the service vehicles. The LSS would also include the Earth Entry Module , an Apollo CM-style conical capsule. In addition to carrying the astronauts through Earth's atmosphere at voyage' s end, it would serve as emergency abort vehic le during the M-1 maneuver. The service vehic les would eac h carry a spare Earth Entry Module.

On the service ships, a hangar for robot probes would replace the LSS . Unlike the Loc kheed and Aeronutronic reports, the General Dynamics report treated its automated Mars probes in some detail. They would include the Returner Mars sample collector, a Mars Lander based on technology developed for NASA's planned Surveyor lunar soft-landing probes , Deimos Probe (Deipro) and Phobos Probe (Phopro) Mars moon hard landers based on tec hnology developed for NASA 's Ranger lunar probes , the Mars Environmental Satellite (Marens) orbiter , and Floater balloons.[13]

General Dynamics' EMPIRE statement of work specified that it should study piloted Mars-orbital missions; however, enthusiastic Ehricke could not resist inserting an option to carry a small piloted Mars lander. A piloted Mars orbiter must, after all, enter and depart Mars orbit, thus performing all the major maneuvers required of a Mars Orbit Rendezvous landing mission except the landing itself The Mars Excursion Vehicle lander, which would be based on the automated Returner, would be carried in a service vehicle probe hangar. It would support two people for seven da ys on Mars.[14] Ehricke's team proposed that a crew test it on the Moon in November 1972.

To get its ships into Earth orbit, Ehricke's team invoked a very large post-Saturn hea vy-lift rocket capable of launc hing 500 tons . Two of these giants would be able to place parts for one ship into orbit so that only one rendezvous and doc king would be required to complete assembly. By contrast, if the Saturn C-5 were used, eight launches and seven rendezvous and doc king maneuvers would be needed to launc h and assemble each General Dynamics Mars ship . The Ehricke team targeted post-Saturn vehicle development to commence in July 1965; the giant rocket would be declared operational in August 1973.

## Mars in Texas

NASA's Manned Spacecraft Center (MSC) (renamed the Johnson Space Center in 1973) began as the Space Task Group (STG) at NASA Langley Researc h Center in Hampton, Virginia, where it w as formed in late 1958 to develop and manage Project Mercury. Following Kennedy's May 1961 Moon speech, the STG's responsibilities expanded, so it needed a new home. The STG became the MSC and moved to Houston, Texas.

Maxime Faget became MSC' s Assistant Director for Research and Development. He launched the first MSC piloted Mars miss ion study in mid-1961, but it remained in-house and at a minimal level of effort until late 1962, after Marshall kic ked off EMPIRE. MSC's study was supervised by Da vid Hammock, Chief of MSC's Spacecraft Technology Division, and Bruce Jackson, one of his branc h chiefs. Chief products of MSC's study were a Mars mission profile unlike any proposed up to that time and the first detailed Mars Excursion Module (MEM) piloted Mars lander design.

Jackson and Hammock presented MSC's Mars plan at the first NASA intercenter meeting focused on interplanetary travel, the Manned Planetary Mission Technology Conference held at Lewis from 21 to 23 May 1963. The NASA Headquarters Office of Applied Research and Technology organized the meeting ,

which focused mainly on specific tec hnologies, many with applications to missions other than Mars . The "Mission Examples" session, chaired by Harry Ruppe, was relegated to the afternoon session on the last da y of the meeting.

Hammock and J ackson presented MSC' s mission design publicly for the first time at the American Astronautical Society (AAS) Symposium on the Manned Exploration of Mars in Denver, Colorado, the first non-NASA conference devoted to piloted Mars travel.[15] George Morgenthaler of Martin Marietta Corporation in Denver organized the symposium. As many as 800 engineers and scientists heard 26 papers and a banquet address by Secretary of the Air Force Eugene Zuckert. It was the first time so many individuals from Mars-related disciplines came together in one place, and the last Mars conference as large until the 1980s. Sky & Telescope magazine reported that the "Denver symposium . . . helped narrow the gaps between engineer, biologist, and astronomer."[16]

Hammock and J ackson called Mars "perhaps the most exciting target for space exploration following Apollo . . . because of the possibility of life on its surface and the ease with whic h men might be supported there."[17] Two of their plans used variations on the MOR mode, but the third, dubbed the Flyby-Rendezvous mode, was novel—it would accomplish a piloted Mars landing while still accruing the weight-minimization benefits of a Crocco-type flyby.

The Flyby-Rendezvous mode would use two separate spacecraft, designated Direct and Flyby . They would reach Earth orbit atop Saturn V rockets. The unpiloted Flyby craft would depart Earth orbit 50 to 100 da ys ahead of the piloted Direct craft on a 200-da y trip to Mars. The Direct craft, which would include the MEM lander, would reach Mars ahead of the Flyby craft after a 120-day flight. The astronauts would then board the MEM and abandon the Direct craft. The MEM would land while the Direct craft flew past Mars into solar orbit. Forty days later the Flyby craft would pass Mars and begin the voyage back to Earth. The crew would lift off in the MEM ascent vehic le and set out in pursuit, boarding the Flyby craft about two days after leaving Mars. Near Earth the astronauts would separate from the Flyby spacecraft in an Earth-return capsule , enter Earth's atmosphere, and land.

One of MSC's MOR plans used aerobraking, while the other relied on propulsive braking. In aerobraking, the lifting-body-shaped Mars spacecraft would skim through Mars' upper atmosphere to use drag to slow down and enter orbit. The Mars surface explorers would separate from the orbiting ship in the MEM and land for a surface stay of 10 to 40 days. They would then lift off in the MEM ascent stage, dock with the orbiting ship, and leave Mars orbit. Earth atmosphere reentry would occur as in the Flyby-Rendezvous mode. Hammock and Jackson's propulsive-braking MOR mission resembled the aerodynamic-braking mode design, except that a chemical or nuclear propulsion stage would place the ship in Mars orbit.

Hammock and Jackson found that the chemical all-propulsive spacecraft design would weigh the most at Earth-orbit departure (1,250 tons), while the nuclear aerobraking design would weigh the least (300 tons). The Flyby-Rendezvous chemical and aerobraking chemical designs would weigh about the same (1,000 tons).

The MEM design for the Houston Center's MOR plans—the first detailed design for a piloted Mars lander—was presented in June 1964 at the next major meeting devoted to Mars exploration, the Symposium on Manned Planetary Missions at Marshall.[18] Philco (formerly Ford) Aeronutronic performed the study between May and December 1963. Franklin Dixon, the presenter, was Aeronutronic's manager for Advanced Space Systems. The design, which the company believed could land on Mars in 1975, was first described publicly in Houston in November 1964 at the American Institute of Astronautics and Aeronautics (AIAA) 3rd Manned Space Flight Conference.

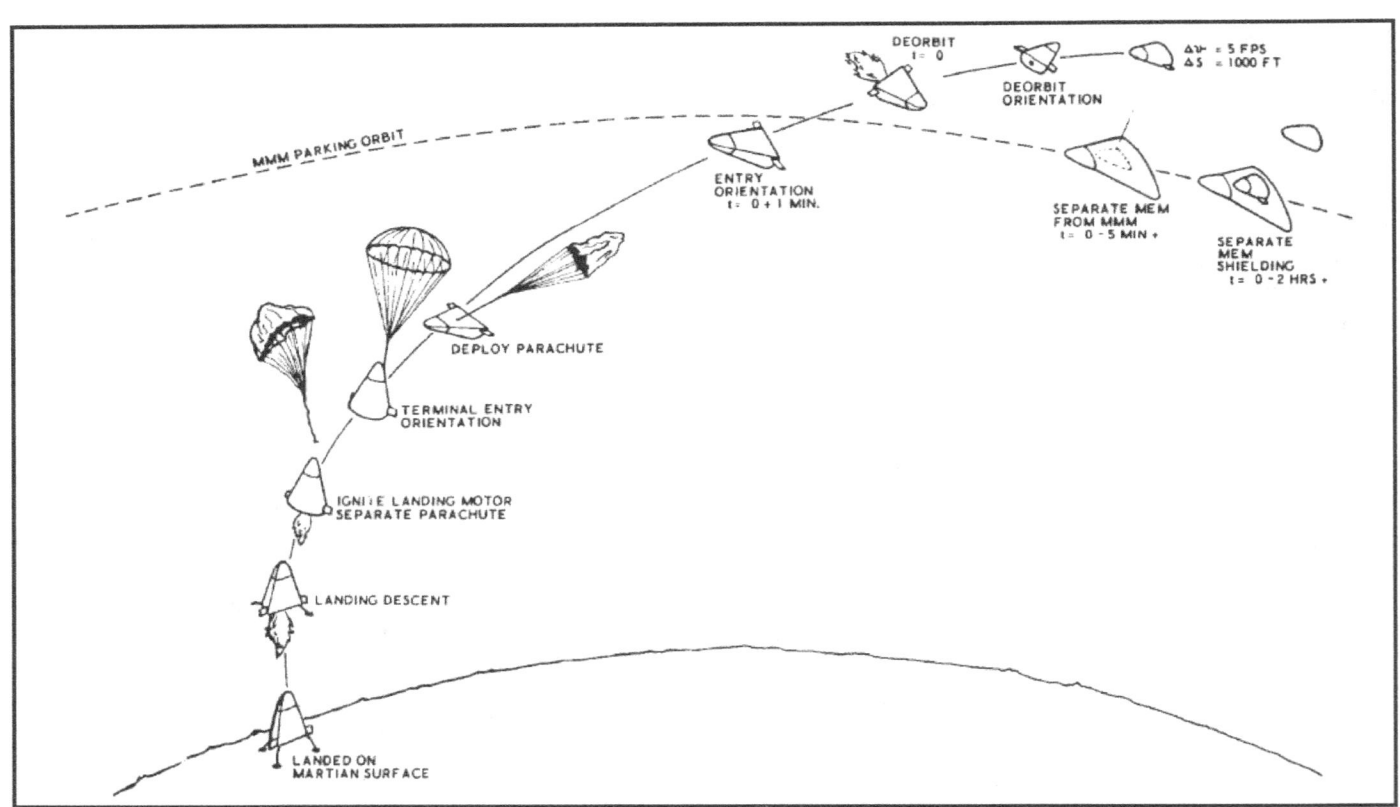

Figure 1—Landing on Mars. Aeronutronic's Mars lander, a lifting body glider, relied on aerodynamic lift to minimize required propellant. The design was based on optimistic estimates of Martian atmospheric density. ("Summary Presentation: Study of a Manned Mars Excursion Module," Franklin Dixon, Proceeding of the Symposium on Manned Planetary Missions: 1963/1964 Status, NASA TM X-53049, Future Projects Office, NASA George C. Marshall Spaceflight Center, Huntsville, Alabama, June 12, 1964, p. 467.)

Dixon pointed out that the chief problem facing Mars lander designers was the lack of reliable Mars atmosphere data, noting that "two orders of magnitude variations in density at a given altitude were possible when comparing Mars atmosphere models of responsible investigators." Aeronutronic settled on a Martian atmosphere comprising 94 percent nitrogen, 2 percent carbon dioxide, 4 percent argon, and traces of oxygen and water vapor, with a surface pressure of 85 millibars (about 10 percent of Earth sea-level pressure). For operation in this atmosphere, Aeronutronic proposed a "modified half-cone" lifting body with two stubby winglets. The Aeronutronic MEM would measure about 30 feet long and 33 feet wide across its tail. The 30-ton MEM would ride to Mars on its mothership's back under a thermal/meteoroid shield which the crew would eject two hours before the Mars landing. The three-person landing party, which would consist of the captain/scientific aide, first officer/geologist, and second officer/biologist, would don space suits and enter the small flight cabin in the MEM's nose. Five minutes before planned deorbit, the MEM would separate from its mothership and retreat to a distance of 1,000 feet. There it would point its tail forward and fire its single descent engine to begin the fall toward Mars' surface.

The MEM's heat-resistant hull would be made largely from columbium, with nickel-alloy aft surfaces. Aeronutronic calculated that friction heating would drive nose temperature to 3,050 degrees Fahrenheit. At Mach 1.5, between 75,000 and 100,000 feet above Mars, a single parachute would be deployed and the MEM would assume a tail-down attitude. The engine would then ignite a second time and the parachute would separate. Aeronutronic's design included enough propellant for an estimated 60 seconds of hover before touchdown on four landing legs with crushable pads.

Aeronutronic attempted to select a MEM landing site using photographs taken by Earth-based telescopes. Theorizing that living things might follow the retreating edge of the melting polar cap in springtime, they suggested that NASA target the MEM to Cecropia at 65 degrees north latitude (this corresponds to Vastitas Borealis north of Antoniadi crater on modern Mars maps).[19] Upon landing, the astronauts would eject shields covering the MEM windows and look out over

Figure 2—Astronauts exploring Mars near Aeronutronic's lander would take pains to collect biological specimens before terrestrial contamination made study impossible. A large dish antenna (left) would let them share their discoveries with Earth. ("Summary Presentation: Study of a Manned Mars Excursion Module," Franklin Dixon, Proceeding of the Symposium on Manned Planetary Missions: 1963/1964 Status, NASA TM X-53049, Future Projects Office, NASA George C. Marshall Spaceflight Center, Huntsville, Alabama, June 12, 1964, p. 470.)

their landing site to evaluate "local hazards," including any "unfriendly life forms."[20] Mars surface access would be through a cylindrical airlock that lowered like an elevator from the MEM's tail.

Dixon stated that "biological evaluation of life forms is essential for the first purely scientific effort to allow pre-contamination studies before man alters the Mars environment,"[21] implying that little effort would be made to prevent the astronauts from introducing terrestrial microorganisms. Aeronutronic listed "investigate life forms for possible nutritional value"[22] among the tasks of the Mars biology study program. The crew would explore Mars for between 10 and 40 days, spending about 16 man-hours outside the MEM each day.

Aeronutronic's MEM was envisioned as a two-stage vehicle. For return to Mars orbit, the ascent motor would fire, blasting the flight cabin free of the descent stage. Two propellant tanks would be cast off during ascent. After docking with the orbiting mothership, the MEM flight cabin would be discarded.

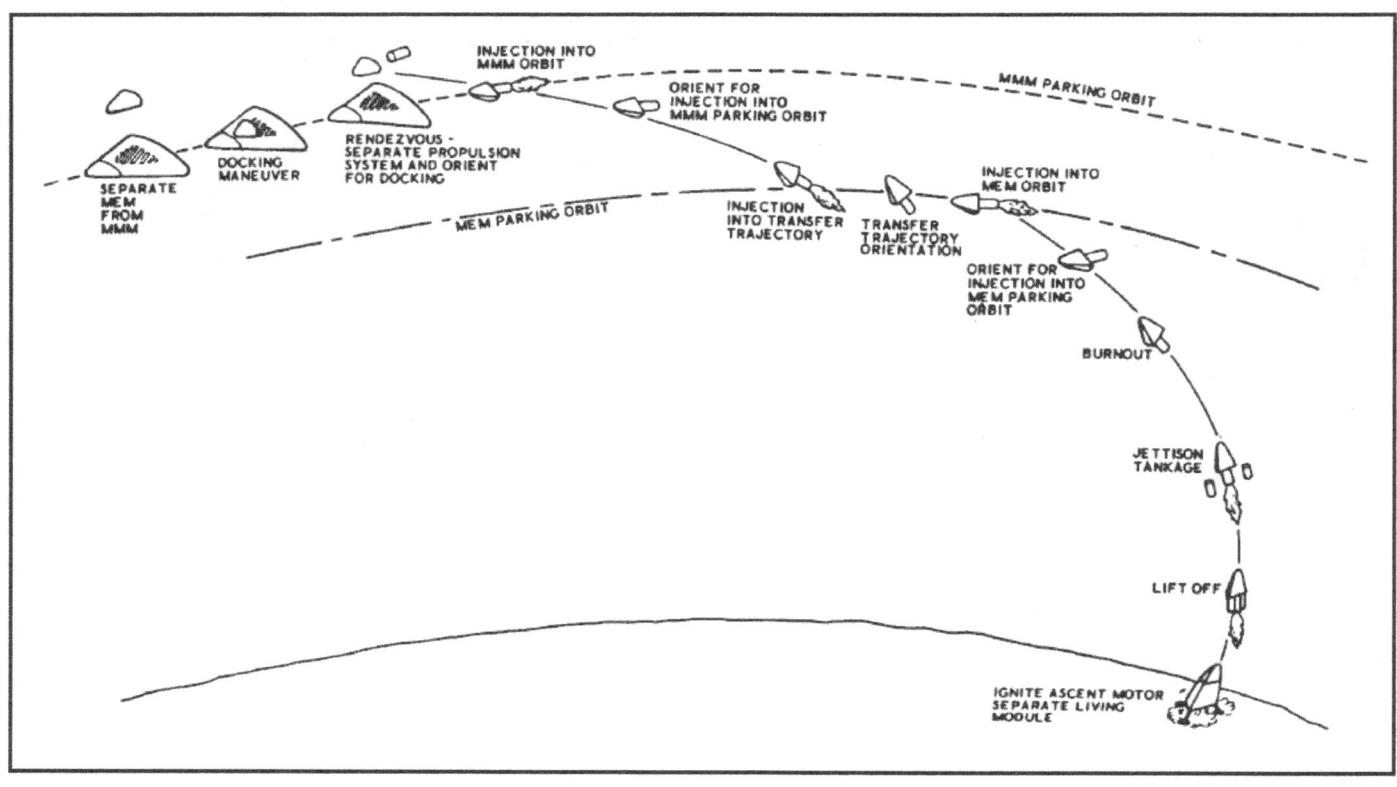

Figure 3—Returning to Mars orbit: Like the Apollo Lunar Module, Aeronutronic's lander design used its descent stage as a launch pad for its ascent stage. Unlike the Lunar Module, it cast off spent propellant tanks as it climbed to orbit. ("Summary Presentation: Study of a Manned Mar s Excursion Module," Franklin Dixon, Proceeding of the Symposium on Manned Planetary Missions: 1963/1964 Status, NASA TM X-53049, Future Projects Office, NASA George C. Marshall Spaceflight Center, Huntsville, Alabama, June 12, 1964, p. 468.)

## UMPIRE

Every 26 months, an opportunity occurs for a short (six-month) minimum-energy transfer from Earth to Mars. In some opportunities the planet is farther from Earth than in others. This means that in some opportunities the minimum energy necessary to reac h Mars is greater than in others. The most difficult Mars opportunities require about 60 percent more energy than the best opportunities. The more energy required to reach Mars, the more propellant a spacecraft must expend. Because of this, a spacecraft launched in a poor Mars opportunity will weigh more than twice as much as one launched in a good Mars opportunity.

The quality of Mars launc h opportunities runs through a continuous cycle lasting about 15 years. Not surprisingly, this corresponds to the cycle of astronomically favorable oppositions described in Chapter 1. The EMPIRE studies showed that the best Mars

opportunities since 1956 would occur in 1969 and 1971, just as the Apollo lunar goal w as reached. Opportunities would become steadily worse after that, hitting a peak in 1975 and 1977, then would gradually improve. The next set of favorable oppositions would occur in 1984, 1986, and 1988.

The Marshall Future Projects Office contracted with General Dynamics/Fort Worth and Douglas Aircraft Company in J une 1963 to "survey all the attractive mission profiles for manned Mars missions during the 1975-1985 time period, and to select the mission profiles of primary interest. " The study, nicknamed "UMPIRE" ("U" stood for "unfavorable"), was summed up in a Future Projects Office internal report in September 1964.[23]

General Dynamics and Douglas worked independently, but each found that the "best method of alleviating the cyclic variation of weight required in Earth orbit is to

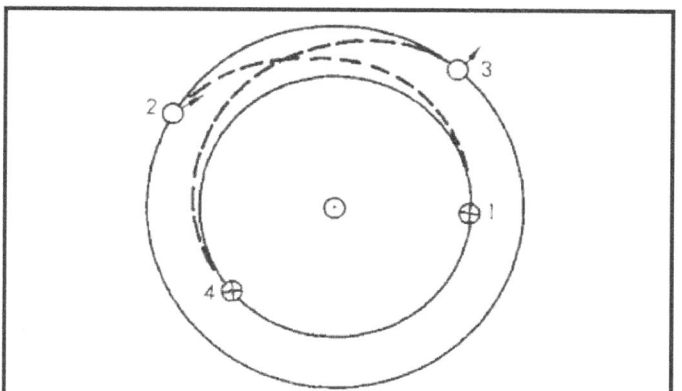

Figure 4—Conjunction-class Mars missions include a low-energy transfer from Earth to Mars, a long stay at Mars, and a low-energy transfer from Mars to Earth. 1 - Earth departure. 2- Mars arrival. 3 - Mars departure. 4 - Earth arrival. (Manned Exploration Requirements and Considerations, Advanced Studies Office, Engineering and Development Directorate, NASA Manned Spacecraft Center, Houston, Texas, February 1971, p. 1-7.)

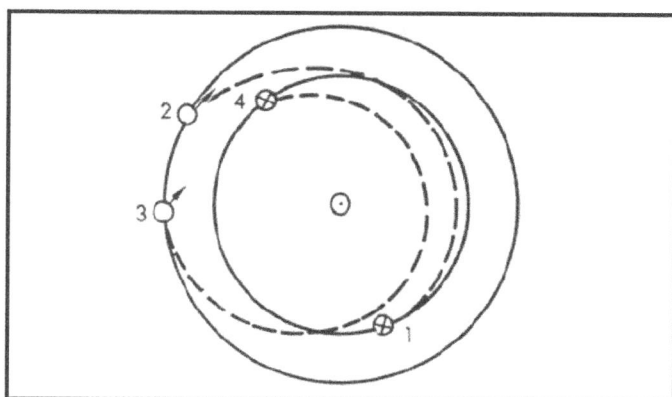

Figure 5—Opposition-class Mars missions offer a short Mars stay but require one high-energy transfer, so they demand more propellant than conjunction-class missions. 1 - Earth departure (low-energy transfer). 2 - Mars arrival. 3 - Mars departure (high-energy transfer). 4 - Earth arrival. (Manned Exploration Requirements and Considerations, Advanced Studies Office, Engineering and Development Directorate , NASA Manned Spacecraft Center, Houston, Texas, February 1971, p. 1-8.)

plan long (900-1100 da ys) missions."[24] The companies advised that "serious consideration . . . be given to the concept of the first manned landing on Mars being a long term base" rather than a short visit.[25] That is, the two companies recommended making the first Mars expedition conjunction class, not opposition class.

The terms "conjunction class" and "opposition class" refer to the position of Mars relative to Earth during the Mars expedition. In the former, Mars moves behind the Sun as seen from Earth (that is, it reaches conjunction) halfway through the expedition; in the latter, Mars is opposite the Sun in Earth' s skies (that is , at opposition) at the expedition's halfway point.

Conjunction-class expeditions are typified by low-energy transfers to and from Mars, each lasting about six months, and by long stays at Mars—roughly 500 days. Total expedition duration thus totals about 1,000 days. The long stay gives Mars and Earth time to reach relative positions that make a minimum-energy transfer from Mars to Earth possible. Von Braun opted for a conjunction-class expedition in The Mars Project.

Opposition-class Mars expeditions ha ve one low-energy transfer and one high-energy transfer separated by a short stay at Mars—typically less than 30 da ys. Total duration is about 600 da ys. This was the approac h

Lewis used in its 1959-1961 study . In the 1960s, most Mars expedition plans were opposition class.

Because they require more energy , opposition-class expeditions demand more propellant. All else being equal, a purely propulsive opposition-class Mars expedition can need more than 10 times as much propellant as a purely propulsive conjunction-c lass expedition. This adds up, of course, to a correspondingly greater spacecraft weight at Earth-orbit departure.

Therefore, the conjunction-class plan is attractive . However, the long mission duration is problematical, for it demands great endurance and reliability from both machines and astronauts, exposes any crew left in Mars orbit to risk from meteoroids and radiation for a longer period, and requires complex Mars surface and orbital science programs to enable productive use of the 500-day Mars stay.

## Mars in California

NASA's Ames Research Center, a former NACA laboratory in Mountainview, California, also became involved in piloted Mars planning in the EMPIRE era. In 1963, Ames contracted with the TRW Space Technology Laboratory to perform a non-nuclear Mars landing expedition study emphasizing weight reduction. Robert Sohn supervised the study for TRW and presented the study's results at the 1964 Huntsville meeting .[26] Sohn's team targeted 1975 for the first piloted Mars landing.

TRW found that the biggest potential weight-saver was aerobraking. For its aerobraking calculations , it used the Rand Corporation' s August 1962 "Conjectural Model III Mars Atmosphere" model, which posited a Martian atmosphere consisting of 98.1 percent nitrogen and 1.9 percent carbon dioxide at 10 percent of Earth sea-level pressure. This atmospheric density and composition dictated the spacecraft's proposed shape— a conical nose with dome-shaped tip, cylindrical center section, and skirt-shaped aft section. This shape w as based on an Atlas missile nose cone. The TRW team's two-stage, 12.5-ton MEM would also use the nose-cone shape. All else being equal, a version of TRW's spacecraft for the 1975 Mars launc h opportunity that used braking rockets at Mars and Earth would weigh 3,575 tons, while the company' s aerobraking design would weigh only 715 tons.

Figure 6—TRW's 1964 Mar s ship design, shaped like a missile warhead, sought to minimize required propellant by aerobraking in the Martian atmosphere . This cutaway shows the Mars lander and Earth Return Module inside the spacecraft. ("Summary of Manned Mars Mission Study," Robert Sohn, Proceeding of the Symposium on Manned Planetary Missions: 1963/1964 Status, NASA TM X-53049, Future Projects Office , NASA George C . Marshall Spaceflight Center, Huntsville, Alabama, June 12, 1964, p. 150.)

TRW's Earth aerobraking system w as the Earth Return Module, a slender half-cone lifting body carried inside the main spacecraft. A few da ys before Earth encounter the crew would enter the Earth Return Module and separate from the main spacecraft. The Earth Return Module would enter Earth' s atmosphere as the main spacecraft flew past Earth into solar orbit.

The TRW study proposed a lightweight artificial gra v ity system—a 500-foot tether linking the main spacecraft to the expended booster stage that pushed it from Earth orbit—which would, it calculated, add less than 1 percent to overall spacecraft weight. The resultant assemblage would spin end over end to produce artificial gravity. TRW reported that NASA Langley had used computer modeling to confirm this design' s long-term rotational stability.[27]

TRW found that Earth-Mars trajectories designed to reduce spacecraft weight at Earth departure would result in high reentry speeds at Earth return. For example, an Earth Return Module would reenter at 66,500 feet per second at the end of a 1975 Mars voyage, while one returning after a 1980 mission would reenter at almost 70,000 feet per second. TRW found that available models for predicting atmospheric friction temperatures broke down at suc h speeds.[28] For comparison, maximum Apollo lunar-return speed w as "only" 35,000 feet per second.

Reentry speed could be reduced by using roc kets. TRW found, however, that including enough propellant to slow the entire spacecraft from 66,500 feet per second to 60,000 feet per second would boost spacecraft weight from 715 tons to 885 tons . Slowing only the Earth Return Module by the same amount would increase overall spacecraft weight to 805 tons .

The study proposed a new alternative—a Venus swingby at the cost of a modest increase in trip time. A ship returning from Mars in 1975 could,the study found, cut its Earth reentry speed to 46,000 feet per second by passing 3,300 kilometers over Venus's night side . A Venus swingby during flight to Mars in 1973 would allow the ship to gain speed without using propellant and thus arrive at Mars in time to take advantage of a slower Mars-Earth return trajectory . According to TRW's calculations, Venus swingby opportunities occurred at every Mars launch opportunity.[29]

## Building on Apollo

By the end of the June 1964 Marshall Mars sympo-sium, early flyby detractor Maxime Faget had come to see some merit in the concept. In a panel discus-sion chaired by Heinz Koelle, Faget declared that "we should, I think, consider a flyby . . . if we under-take a flyby we really have to face the problems of man flying out to interplanetary distances . . . . I think we have to undertake a program that will force the technology, otherwise we will not get [to Mars] in my lifetime . . . ."[30]

Von Braun, also a panel member, added, that "I think [piloted] flyby missions, particularly flybys involving [automated] landing probes . . . would be invaluable . . . . One such flight, giving us more information on what to expect . . . on the surface of Mars, will be extremely valuable in helping us in laying out the equipment for the landing . . . that would follow the first flyby flight."[31]

Von Braun then implicitly announced an impending shift in NASA advanced planning. "I am also inclined to believe," he said, "that our first manned planetary flyby missions should be based on the Saturn V as the basic Earth-to-orbit carrier. The reason is that, once the pro-duction of this vehicle is established and a certain reli-ability record has been built up, this will be a vehicle that will be rather easy to get." Von Braun's statement acknowledged that a post-Saturn rocket appeared increasingly unlikely.[32] In an outline of future plans submitted to President Lyndon Johnson's Budget Bureau in late November 1964, NASA stated that the post-Saturn rocket should receive low funding priority, and called for post-Apollo piloted spaceflight to be focused on Earth-orbital operations using technology developed for the Apollo lunar landing.[33]

The 1964 decision to use Apollo technology for missions after the lunar landing could be seen as a rejection of post-Apollo piloted Mars missions. Historian Edward Ezell wrote in 1979 that the "determinism to utilize Apollo equipment for the near future was very destruc-tive to the dreams of those who wanted to send men to Mars."[34] As if to emphasize this, the amount of funding applied to piloted planetary mission studies took a nose dive after November 1964. In the 17 months preceding November 1964, $3.5 million was spent on 29 piloted planetary mission studies. Between November 1964

and May 1966, NASA contracted for only four such studies at a cost of $465,000.[35]

Mars planners were not so easily discouraged, however. After EMPIRE, and concurrent with UMPIRE, a Marshall Future Projects Office team led by Ruppe com-menced an in-house study to look at using Apollo hard-ware for Mars exploration. Ruppe's study report, pub-lished in February 1965, found that piloted Mars flyby missions would be technically feasible in the mid- to late-1970s using Saturn rockets and other Apollo hardware.[36] The report's flyby spacecraft design used hardware already available or in an advanced state of development. Two RL-10 engines would provide rendezvous and dock-ing propulsion, for example, and an Apollo Lunar Module descent engine would perform course corrections.

A pressurized hangar would protect a modified Apollo CSM during the interplanetary voyage. The hangar would also provide a shirt-sleeve environment so that the astronauts could act as in-flight caretakers for five tons of automated probes, including "landers, atmospheric floaters, skippers, orbiters, and possibly probes . . . to perform aerodynamic entry tests [of] designs and materials."[37] The last of these would, Ruppe wrote, provide data to help engineers design the piloted Mars landers to follow. His report drew on the UMPIRE conclusions when it stated that

significant reduction of initial mass in Earth orbit is possible if we can use aerodynamic braking at Mars or refueling there, but these methods assume a knowledge about . . . the Martian atmosphere, or about Mars surface resources which just is not available. The first venture, still assuming that we are not very knowledgeable . . . would probably transport 2 or 3 men to the surface of Mars for a few days . . . [at a cost of] a billion dollars per man-day on Mars. If the physical properties of Mars were well known, we could think . . . of the first landing as a long-duration base, reducing cost to less than 10 million dollars per man-day.[38]

The three-person flyby crew would live in a spherical habitat containing a radiation shelter and a small cen-trifuge for maintaining crew health (the study rejected artificial gravity systems that rotated the entire craft as being too complex and heavy). Twin radioisotope power units on extendible booms would provide electricity.

The mission would require six Saturn V launches and one Saturn IB launch. Saturn V rocket 1 would launch the unpiloted flyby spacecraft; then Saturn V rockets 2 through 5 would launc h liquid oxygen tankers . The sixth Saturn V would then launch the Earth-departure booster, a modified Saturn V second stage called the S-IIB, which would reach orbit with a full load of 80 tons of liquid hydrogen but with an empty liquid oxygen tank. Ruppe wrote that solar heating would cause the liquid hydrogen to turn to gas and escape; to ensure that enough remained to boost the flyby craft tow ard Mars, the S-IIB would have to be used within 72 hours of launch from Earth.

The three astronauts would launc h in the modified Apollo CSM on the Saturn IB roc ket and then board the flyby spacecraft. They would use the RL-10 engines to guide the flyby craft to a doc king with the S-IIB. The oxygen tankers would then dock in turn and pump their cargoes into the S-IIB' s empty oxygen tank. Ruppe's flyby craft and booster would weigh 115 tons at Earth-orbit departure . The S-IIB would then ignite, burn to depletion, and detach, placing the flyby craft on course for Mars.

During the flight, the astronauts would regularly inspect and service the automated probes . As they approached Mars, the astronauts would release the probes and observe the planet using 1,000 pounds of scientific equipment. The flyby spacecraft would rela y radio signals at a high data rate between the Mars probes and Earth until it passed out of range; then direct communication between Earth and the probes would commence at a reduced data rate .

As Earth grew large again outside the viewports , the flyby astronauts would enter the modified Apollo CSM and abandon the flyby craft. The CSM's propulsion system would slow it to Apollo lunar return speed, then the CM would separate from the SM, reenter, and land. Depending on the launch opportunity used, total mission duration would range from 661 to 691 da ys.

Even as Ruppe's report was published, the "robot care-taker" justification for piloted Mars flybys w as becoming increasingly untenable . On 31 J uly 1964, the Ranger 7 Moon probe snapped 4,316 images of one corner of Mare Nubium before smashing into the lunar surface as planned. The images showed the Moon to be sufficiently smooth for Apollo landings, and gave the credibility of robot explorers a vital boost. As Ruppe published his report, Mariner 4, launched on 28 November 1964, was making its way toward Mars. Not long after Ruppe published his report, on 20 February 1965, Ranger 8 returned 7,137 images as it plunged toward the Sea of Tranquillity. A month later, Ranger 9 returned 5,148 breathtaking images of the complex 112-kilometer crater Alphonsus.

Beyond providing engineering and scientific justifications for the piloted flyby mission, Ruppe's report tendered a political justification. He wrote:

> From the lunar landing in this decade to a possible planetary landing in the early or middle 1980s is 10 to 15 years . Without a major new undertaking , public support will decline. But by planning a manned planetary [flyby] mission in this period . . . the United States will stay in the game.[39]

That Ruppe felt it necessary in early 1965 to attempt to justify a piloted Mars flyby mission in terms of probable impact on the U.S. domestic political environment is telling, as will be seen in the next c hapter.

# Chapter 4: A Hostile Environment

An era ended for the National Aeronautics and Space Administration last week when Congress voted a $234-million cut in that agency's budget authorization for Fiscal 1968 . . . . The NASA budget cut is symptomatic of the many currents of basic change that are flowing through the land this summer . . . . If top NASA officials have not interpreted their admittedly long and arduous buffeting on Capitol Hill this spring and summer correctly, then they are facing a much worse time in the years ahead . . . . (Robert Hotz, 1967)[1]

## Mariner 4

On 15 July 1965, the Mariner 4 probe snapped 21 blurry pictures of Mars' southern hemisphere as it flew by at a distance of 9,600 kilometers . The flyby, which marked the culmination of a seven-and-a-half-month voyage, was an unprecedented engineering achievement. Mariner 4 had withstood the interplanetary environment for nearly twice as long as Mariner 2 had during its 1962 Venus flyby mission.

Mariner 4 revealed Mars to be a disappointingly Moonlike, cratered world with no obvious signs of water. Scientists had expected to see a world more like Earth, where erosion makes obvious craters the exception rather than the rule. That Mariner 4's images were black and white accentuated the resemblance to Earth's desolate satellite. Canals were conspicuously absent. They are now believed to have been an optical illusion or a product of eyestrain.

Mariner 4's impact on Mars exploration planning is hard to overestimate. First, it showed that Maxime Faget had been right in 1962. Robots could perform Mars flybys—astronauts were not required for this particular exploration mission. It also showed that robot probes could reach Mars in reasonably good condition, undermining the "robot caretaker" justification for piloted Mars flybys.

Mariner 4's radio-occultation experiment revealed Mars' atmosphere to be less than 1 percent as dense as Earth's. Based on these new data and on measurements of the Martian atmosphere made from Earth since the 1940s, planetary scientists calculated that the majority of Mars' atmosphere was carbon dioxide, not nitrogen, as had been widely supposed.[2]

The new Mars atmosphere data relegated to the recycle bin aerodynamic landing systems such as von Braun's delta-winged gliders and Aeronutronic's lifting-body. That meant more rocket propulsion would be required to accomplish a Mars soft landing, which would in turn demand more propellant. This would boost minimum lander weight, which meant more propellant would be needed to transport the lander from Earth to Mars. This in turn would boost Mars spacecraft weight at Earth-orbit departure, which meant, of course, that more expensive rockets would be required to launch the Mars ship into Earth orbit.

Most importantly, Mariner 4 dealt a body blow to hopes for advanced Martian life . Historically, human perceptions of life on Mars have occurred along a continuum. At one end stood the romantic view of nineteenth-century American astronomer Percival Lowell, whose Mars was a dying Earth inhabited by a race of civil engineers who had dug a planet-girdling network of irrigation canals to stave off the encroaching red desert. By the 1930s, Lowell's vision was widely seen as optimistic . Nonetheless, the romance of Lowell's Mars inspired would-be Mars explorers into the 1960s.[3]

The Mariner 4 results eradicated any lingering traces of Lowellian romance, and in fact shifted the prevailing view of life on Mars all the way down the continuum to a pessimism with almost as little basis as Lowell's optimism. The spacecraft had, after all, imaged only 1 percent of Mars at resolution so low that, had it photographed Earth, scientists examining its pictures would likely have missed all signs of terrestrial life.[4] NASA took pains to point out that Mariner 4 had been intended only as a first, preliminary step toward resolving the question of life on Mars, and that it had "blazed the way for later spacecraft to land instruments and, eventually, men on Mars."[5]

On the plus side , Mariner 4 provided the first firm data on conditions astronauts could expect to encounter in interplanetary space during the voyage to Mars. The intrepid robot registered fewer meteoroid impacts than expected, but also detected a higher-than-expected level of cosmic radiation and between 12 and 20 solar flares during what was expected to be a quiet Sun period.[6]

# Chapter 4: A Hostile Environment

## Vietnam and Watts

President Lyndon Johnson supported the lunar program launched by his predecessor, which was not surprising, given that he had pla yed a key role in formulating the Moon goal as K ennedy's Vice President and National Space Council c hair. Like many others, however, he was uncertain what NASA's scope and direction should be in the years after it put an astronaut on the Moon. In a letter on 30 J anuary 1964, Johnson asked NASA Administrator James Webb for a list of possible future NASA goals.[7]

As stated in the last c hapter, the outline of agency plans submitted to J ohnson's Budget Bureau in November 1964 emphasized using Apollo hardware in Earth orbit. An Apollo-based piloted program in the early 1970s was seen as an interim step to an Earth-orbiting space station in the mid- to late-1970s .[8] When the National Academy of Sciences Space Science Board called instead for an emphasis on planetary exploration, NASA officials insisted that the Earth-orbital focus was President Johnson's preference.[9]

This philosophy—that the United States would be best served by using Apollo hardware as an interim step to a future space station—set the tone for much of NASA's post-Apollo planning through the beginning of 1969. NASA's program for reapplying Apollo hardware was the Apollo Applications Program (AAP), an initially ambitious slate of lunar and Earth-orbital missions that eventually shrank to become the Skylab program. As shown in the last c hapter, Mars planners in the Future Projects Office at Marshall sought also to apply Apollo technology to Mars exploration.

An event on 25 J anuary 1965 also helped set the tone for NASA's post-Apollo future. On that date, President Johnson sent to Congress a $5.26-billion NASA budget for FY 1966, an increase of only $10 million over the $5.25-billion FY 1965 budget. This was the smallest NASA budget increase since the agency w as established in 1958. NASA's eventual FY 1966 appropriation was $5.18 billion, the Agency's first budget drop. Most of the cuts came from AAP and other new starts.

This new frugality in the administration and in Congress with regards to space reflected growing unease across the United States . In August 1964, following a naval incident in the Gulf of Tonkin off North Vietnam, Congress passed the Tonkin Resolution, which empowered President J ohnson to take what steps he deemed necessary to thw art further communist aggression in Indoc hina. In February 1965, Vietcong guerrillas attacked the South Vietnamese military base at Pleiku, killing 8 Americans and wounding 126. In response, Johnson ordered the bombing of North Vietnam's base at Dong Hoi. On 8 Marc h, the first U.S. combat troops—two battalions of marines—joined the 23,000 American advisors already in South Vietnam.

As Mariner 4 approached Mars in J uly, President Johnson announced that he would increase the number of soldiers in South Vietnam from 75,000 to 125,000. On 4 August, while Mariner 4's images were trickling back to Earth, Johnson asked Congress for an additional $1.7 billion to support the expanding w ar.

On 11 August, as Mars planners attempted to reconcile the thin atmosphere and craters revealed by Mariner 4 with their old plans for Mars, racial violence flared in the Watts ghetto of Los Angeles, California. Five nights of anarchy left 34 dead and caused $40 million in damage .

## Planetary JAG

Against this bac kdrop of w ar, social unrest, and Mariner 4 results, NASA launched a two-prong assault on Mars. The first, the Voyager program, aimed at planetary exploration using automated orbiters and landers. The second w as an internal piloted Mars flyby study involving several NASA centers.

As already indicated, planetary scientists had rejected the AAP space station emphasis in fa vor of planetary exploration, which, they felt, was being neglected in NASA's headlong rush to reac h the Moon. In its report Space Research: Directions for the Future, released in January 1966, the National Academy of Sciences Space Science Board designated "the exploration of the near planets as the most rew arding goal on whic h to focus national attention for the 10 to 15 years following manned lunar landing."[10] In May 1966, the American Astronomical Society Symposium "The Search for Extraterrestrial Life" re-emphasized the importance of seeking life on Mars despite the Mariner 4 results .[11] These inputs helped build both Voyager and piloted flyby mission rationales.

Voyager, first proposed in 1960 at the J et Propulsion Laboratory (JPL) in P asadena, California, was envisioned as a follow-on program to the Mariner flyby series. The 1960s Voyager should not be confused with the twin Voyager flyby probes launc hed to the outer planets in 1977 and 1978. In the FY 1967 budget cycle, NASA had postponed proposing Voyager as a new start following assurances that it could get off to an aggressive start in FY 1968. The delay was partly a result of the Mariner 4 findings. New atmosphere data forced a re-design that drove the program' s estimated cost beyond $2 billion. [12] Voyager was initially targeted for first launch in 1971, with a second mission in 1973, and other missions to follow.

The NASA Headquarters Office of Manned Space Flight (OMSF) under George E. Mueller, Associate Administrator for Manned Space Flight, managed the piloted flyby study. Mueller had taken c harge of the OMSF in September 1963 and had set up the Advanced Manned Missions Office under Edw ard Gray in November 1963 to direct NASA 's piloted planetary mission planning activities. At a meeting on 15 April 1965, Mueller had received authority from NASA Deputy Administrator Robert Seamans to put together a NASA-wide group to plan piloted planetary missions. A preliminary meeting of the group occurred on 23 April 1965. This prepared the ground for development of the Planetary Joint Action Group (JAG), which was formally established later in the year. The Planetary JAG was headed by Gray and drew members from NASA Headquarters , Marshall, MSC, and Kennedy Space Center (KSC), as well as from the Apollo planning contractor, Bellcomm. [13]

Initially the Planetary J AG's focus w as on piloted Mars missions using nuclear rockets. In April 1966, however, Mueller launched a piloted Mars flyby study within the Planetary J AG at the request of Nobel Laureate Charles Townes, chair of the NASA Science and Technology Advisory Committee. Townes had asked Mueller in January 1966 to carry out a study comparing the unpiloted Voyager project with a piloted flyby with robot probes (what he called a "manned Voyager"). [14] In the second half of 1966, NASA spent $2.32 million on 12 piloted plan-

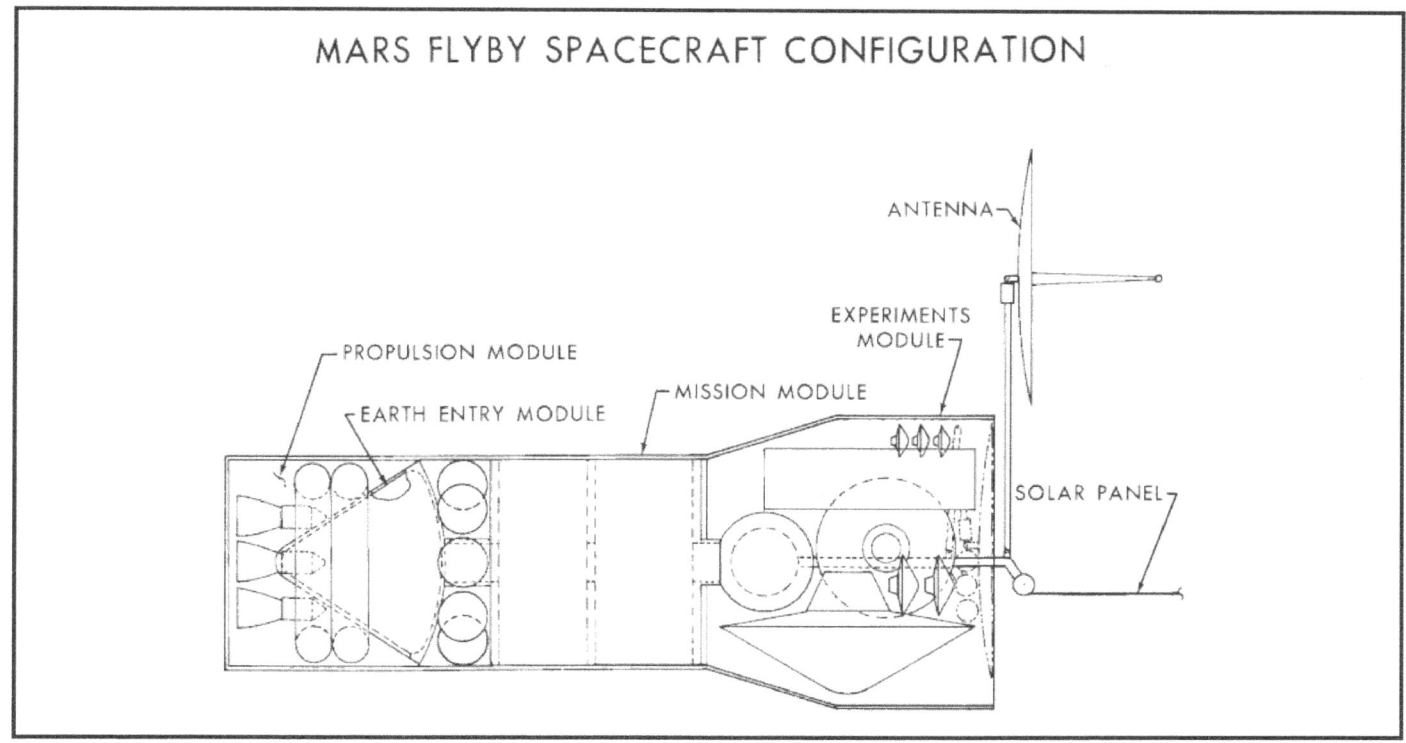

**MARS FLYBY SPACECRAFT CONFIGURATION**

Figure 7—The 1966 Planetary Joint Action Group study used existing and near-term technology for its piloted Mar s flyby spacecraft design. Note the Earth Entry Module (left) based on the Apollo Command Module. (NASA Photo S-66-11230)

etary mission studies supporting the Planetary JAG.[15]

Later that year, Mueller testified to the House Space Committee on the benefits of a piloted flyby . He explained that it afforded

> the best opportunity for performing manned planetary exploration with minimal cost and at an early date . . . . The attractiveness of this type of mission . . . stems from the relatively light burden which it imposes on the propulsion system, although the short interval of direct contact with the target planet detracts from its desirability. The usefulness of the flyby mission becomes c learly established when viewed as an in-situ test-bed for evaluating the performance of various subsystems suc h as navigation, life support, and communications to be used in later landing missions; [and] when also viewed as a platform for launc hing instrumented probes toward the target planet during the close passage.[16]

On 3 October 1966, the Planetary JAG published its Phase 1 report, Planetary Exploration Utilizing a Manned Flight System.[17] The report placed piloted flybys within an evolutionary "integrated program" of new and Apollo-based technology with "balanced" use of humans and robots , the objective of whic h was "maximum return at minimum cost, assuming intensive investigation of the planets is a goal. " By this time the integrated program concept had been discussed for more than a year outside NASA. [18] The Planetary JAG's integrated program proceeded through the following steps:

- Apollo Applications Program (1968-73): Astronauts would remain aloft in space stations based on Apollo hardware for progressively longer periods to collect data on human reactions to weightlessness . Some would live in Earth orbit for more than a year approximately the duration of a piloted Mars flyby mission.

- Mariner (1969-73) and Voyager (1973): The Planetary JAG report cast Mariner and Voyager as lead-ins to piloted expeditions by stating that data they collected would aid engineers designing piloted flyby hardw are. A

Mariner probe would fly by Mars in 1969; in 1971 another Mariner would drop a probe into Mars' atmosphere. The first Voyager probe would land on Mars in 1973 bearing a suite of life-detection experiments.

- Piloted Mars/Venus Flybys (1975-80): The first piloted Mars flyby mission would lea ve Earth-orbit in September 1975. Mars flyby launc h opportunities would also occur in October 1977 and November 1979. Multiple flyby missions were possible—a Venus/Mars mission could start in December 1978, and a Venus/Mars/ Venus mission could launch in February 1977. These would dispense automated probes based on Mariner and Voyager technology.

- Piloted Mars Landin g and piloted Venus Capture (orbiter) missions (post-1980) would see introduction of AEC-NASA nuclear-thermal rockets. The Planetary J AG deemed nuclear propulsion "essential for a flexible Mars landing program" capable of reac hing Mars in any launc h opportunity regardless of the energy required. (The nuclear rocket program is described in more detail in Chapter 5.)

The Planetary JAG's piloted Mars flyby spacecraft would reach Earth orbit on an Improved Saturn V rocket with a modified S-IVB (MS-IVB) third stage. The MS-IVB would feature stretc hed tanks to increase propellant capacity and internal foam insulation to permit a 60-hour wait in Earth orbit before solar heating caused its liquid hydrogen fuel to turn to gas and escape.

The four-person flyby crew would ride into Earth orbit on a two-stage Improved Saturn V in an Apollo CSM stacked on top of the flyby craft. Upon reaching orbit, the CSM/flyby craft combination would detach from the spent Saturn V S-II second stage; then the astronauts would detach the CSM, turn it around, and dock with a temporary docking structure on the flyby craft' s forward end.

The Planetary JAG's flyby spacecraft would consist of the Mid-Course Propulsion Module with four main engines; the Earth Entry Module , a modified Apollo CM for Earth atmosphere reentry at mission' s end; and the Mission Module, the crew's living and working

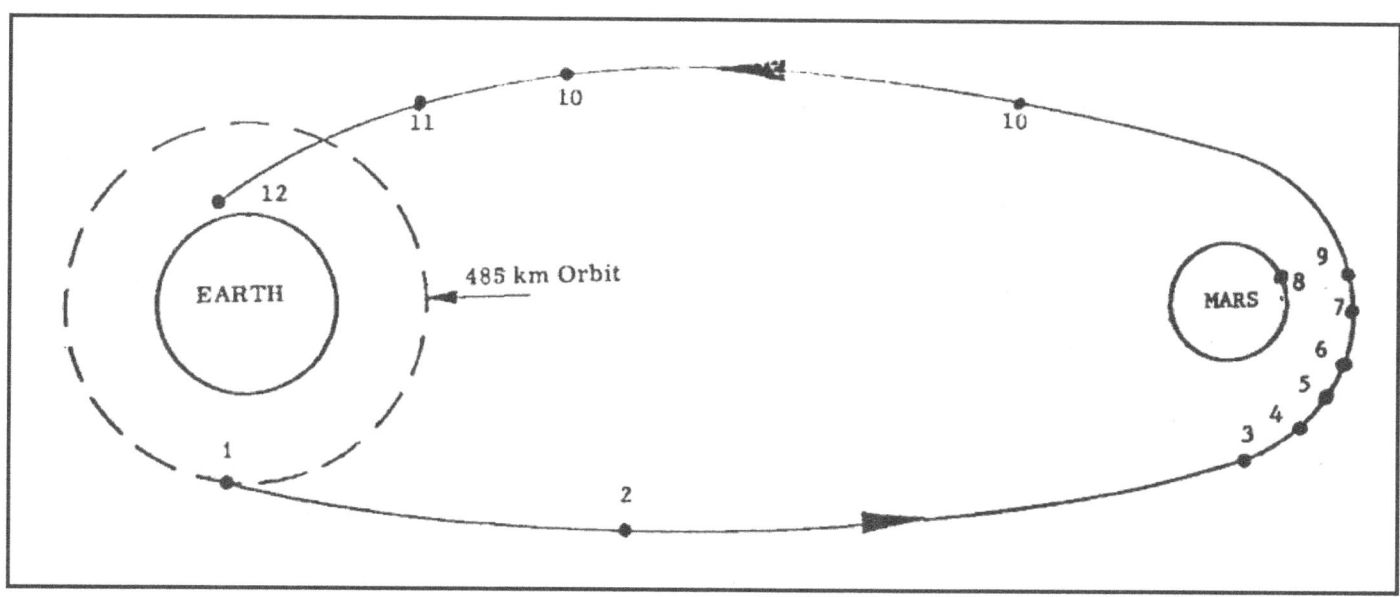

Figure 8—Typical piloted Mars flyby mission. 1—depart Earth orbit. 2, 4, 10—course corrections. 3, 5, 6, 7—eject automated Mars probes. 8—automated probe collects Mars surface sample and launches it off the planet. 9—piloted flyby craft retrieves Mars surface sample. 11—crew leaves Mars flyby craft in Earth return capsule . The abandoned flyby spacecraft sails past Earth into solar orbit. 12—Earth atmosphere reentry and landing. (Planetary Exploration Utilizing a Manned Flight System, Office of Manned Space Flight, NASA Headquarters, Washington, DC, October 3, 1966, p. 16.)

space. The Earth Entry Module would serve double duty as a radiation shelter during solar flares . Mid-Course Propulsion Module propellant tanks would be clustered around it to provide additional radiation shielding. The Mission Module's forward level (for "rest and privacy") would be lined with loc kers containing freeze-dried foods; the aft level would contain the flyby craft's control console, science equipment, and ward-room table. The Planetary JAG report proposed that the Mission Module structure and subsystems, such as life support, be based on Earth-orbital space station module designs.

The automated pr obes would be housed in the Experiment Module forming the aft end of the flyby spacecraft, along with a probe deployment manipulator arm, a biology laboratory, a 40-inch telescope, and an airlock for spacewalks with an Apollo-type docking unit. A 19-foot-diameter radio dish antenna for high-data-rate communications with Earth unfolded from the back of the Experiment Module , as did a 2,000-square-foot solar array capable of generating 22 kilowatts of electricity at Earth, 8.5 kilowatts at Mars, and 4.5 kilowatts in the Asteroid Belt.

With the crew and flyby craft in Earth orbit, three Improved Saturn V rockets would launc h 12 hours apart to place the three MS-IVB rocket stages in orbit.

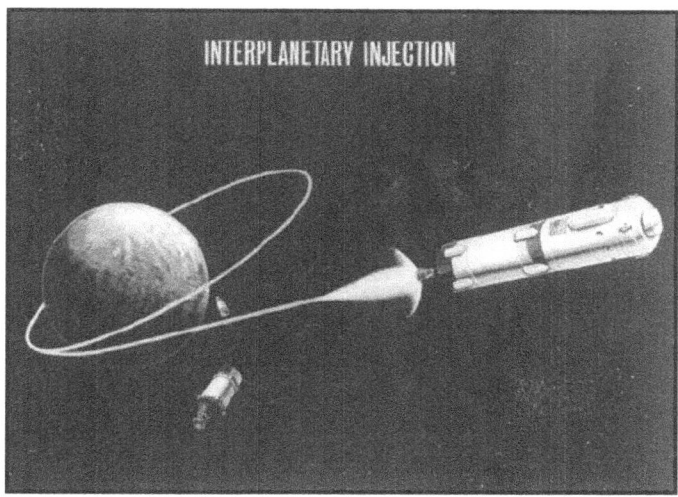

Figure 9—Three modified Apollo S-IVB stages burn one after the other to launch the 1967 Planetary Joint Action Group Mars flyby spacecraft out of Earth orbit. (NASA Photo S-67-5998)

Figure 10—Following final S-IVB stage separation, the 1967 Planetary Joint Action Group's Mars flyby spacecraft deploys solar arrays and a dish-shaped radio antenna.(NASA Photo S-67-5991)

Figure 11—The 1967 Planetary Joint Action Group's Mars flyby spacecraft releases automated probes and deploys instruments. Close Mars flyby would last mere hours, but the astronauts would study themselves throughout the mission, helping to pave the way for future Mars landing expeditions. (NASA Photo S-67-5999)

This rapid launch rate, a veritable salvo of 3,000-ton rockets, each nearly 400 feet tall, would demand construction of a third Saturn V launch pad at KSC. The Planetary JAG determined, however, that Pad 39C would be the only major new ground facility needed to accomplish its flyby program.

Using the CSM's propulsion system, the astronauts would perform a series of rendezvous and docking maneuvers to bring together the flyby craft and three MS-IVBs. The flyby crew would then undock the CSM from the temporary docking structure, re-dock it to the airlock docking unit on the flyby craft's side, and enter the flyby craft for the first time. They would discard the CSM and eject the temporary docking structure.

Launch from Earth orbit would occur between 5 September and 3 October 1975. The MS-IVB stages would in turn ignite, deplete their propellants, and be discarded. As Earth and Moon shrank in the distance, the crew would deploy the radio antenna and rectangular solar array.

The astronauts would perform a wide range of scientific experiments during the 130-day flight to Mars. These included solar studies, monitoring themselves to collect data on the physiological effects of weightless-

ness, planetary and stellar observations, and radio astronomy far from terrestrial radio interference.

Mars flyby would occur between 23 January and 4 February 1976, the precise date being dependent on the date of Earth departure. Beginning several weeks before flyby, the crew would turn the craft's telescope toward Mars and its moons. The pace would quicken 10 days before flyby, when the flyby craft was 2 million kilometers from Mars. At that time the astronauts would use the probe deployment arm to unstow and release the automated probes. At closest approach, the flyby spacecraft would fly within 200 kilometers of the Martian dawn terminator (the line between day and night).

For the 1975 mission, the flyby craft would carry in its Experiment Module three 100-pound Mars impactors, one five-ton Mars polar orbiter, one 1,290-pound Mars lander, and one six-ton Mars Surface Sample Return (MSSR) lander. The MSSR was designed to leave the flyby craft, land on Mars, gather a two-pound sample of dirt and rock, and then blast it back to the passing flyby craft using a three-stage liquid-fueled ascent vehicle.

This last concept, an effort to improve the piloted flyby mission's scientific productivity, was proposed by

Bellcomm at the Planetary JAG's meeting at KSC on 29-30 June 1966.[19] The concept originated in a paper by R. R. Titus presented in January 1966.[20] Titus, a United Aircraft Research Laboratories engineer with a talent for unfortunate acronyms, dubbed his concept FLEM, for "Flyby-Landing Excursion Mode." He had been part of the Lockheed EMPIRE team.

Titus' mission plan had a piloted MEM lander separating from the flyby spacecraft during the Mars voyage and changing course to intersect the planet. Titus calculated that MEM separation 60 days before Mars flyby would permit it to stay for 16 days at Mars, while separation 30 days out yielded a 9-day stay time. At Mars the MEM would fire its rocket engine to enter orbit, then land. As the flyby spacecraft passed Mars, the excursion module would blast off in pursuit. The amount of propellant required for FLEM was much less than for an MOR landing mission because only the MEM would enter and depart Mars orbit. Titus calculated that a FLEM mission boosted to Mars in 1971 using a nuclear-thermal rocket might weigh as little as 130 tons—light enough, perhaps, to permit a piloted Mars landing with a single Saturn V launch.

In the 1966 JAG piloted flyby plan, the automated MSSR would land on Mars about two hours before the flyby craft flew past the planet and would immediately set to work gathering rock and soil samples using scoop, brush, sticky tape, drill, and suction collection devices. Less than two hours after MSSR touchdown, its ascent vehicle first stage would ignite. If all went well, the ascent vehicle's small third stage would deliver the samples to a point in space a few miles ahead of the flyby craft about 17 minutes later, 5 minutes after the flyby craft's closest Mars approach. As their craft overtook the sample package, the astronauts would snatch it in passing using a boom-mounted docking ring. They would then deposit it inside the Experiment Module's biology lab.

The Planetary JAG pointed out that the MSSR/piloted flyby approach improved the chances for studying living Martian organisms because the Mars samples would reach a trained biologist within minutes of collection. Living organisms collected using a purely automated sample-return lander would likely perish during the months-long flight to a lab on Earth.

The trip back to Earth would last 537 days, during which the astronauts would study the Mars samples and repeat many of the same experiments performed during the Earth-Mars voyage. The flyby craft would penetrate the Asteroid Belt before falling back to Earth, making piloted asteroid flybys a possibility. When farthest from the Sun the flyby craft would be on the opposite side of the Sun from Earth, making possible simultaneous observations of both solar hemispheres.

A few days before reaching Earth, the crew would board the Earth Entry Module and abandon the flyby craft. On 18 July 1977, the Earth Entry Module would reenter Earth's atmosphere, deploy parachutes, and lower to a land touchdown, while the flyby craft would fly past Earth into solar orbit. Just before the landing, solid-propellant rocket motors would fire to cushion impact, ending the 667-day Martian odyssey.

## The Fire

NASA's FY 1967 funding request was $5.6 billion. The White House Budget Bureau trimmed this to $5.01 billion out of a $112 billion Federal budget before sending the budget on to Capitol Hill. By the time President Johnson signed it into law, NASA's FY 1967 appropriation was $4.97 billion, more than $200 million less than FY 1966. Programs aimed at giving NASA a post-Apollo future were hardest hit. Of the $270 million NASA requested for AAP, for example, only $83 million was appropriated. Voyager funding start-up was bumped to FY 1968. Apollo Moon program funding, by contrast, barely suffered. In part this was because the agency was flying frequent Gemini missions—10 in 20 months—which kept the Moon goal in the public eye. Kennedy's goal seemed very close, with the first piloted lunar landing expected in just over a year.

In Gemini's last year, however, America's attention was increasingly drawn away from space. In March 1966, protesters marched against the Vietnam War in Boston, San Francisco, Chicago, Philadelphia, and Washington. The summer of 1966 saw race riots in Chicago and Atlanta and racist mob violence in Grenada, Mississippi. In June 1966, President Johnson ordered bombing raids against the North Vietnamese cities of Haiphong and Hanoi. By then, 285,000 Americans were serving in Vietnam. As Gemini 12, the last in the series, splashed down in November 1966, the number of American soldiers on the ground in Vietnam was well on its w ay to its 1 J anuary 1967 total of 380,000.

Against this backdrop, in January 1967, the Planetary JAG resumed piloted flyby planning, this time with the purpose of developing "a clear statement of the activities required in FY 69 for budget discussions"[21] to place NASA "in a position to initiate a flyby project in FY 1969."[22] Planetary JAG participants had some reason to be hopeful. As they reconvened, President Johnson announced a $5.1-billion FY 1968 NASA budget that included $71.5 million for Voyager and $8 million for advanced planning. He also backed $455 million for a substantial AAP. In presenting his budget, the President explained that "we have no alternative unless we wish to abandon the manned space capability we have created."[23]

On 26 January the OMSF presented its ambitious AAP plans to Congress . Barely more than a da y later, NASA's plans received a harsh blow as fire erupted inside the AS-204 Apollo spacecraft on the launc h pad at KSC, killing Apollo 1 astronauts Gus Grissom, Ed White, and Roger Chaffee. They had been scheduled to test the Apollo CSM for 14 da ys in Earth orbit beginning in mid-February. NASA suspended piloted Apollo flights pending the outcome of an investigation. The AS-204 investigation report, issued in April, found shortcomings in Apollo management, design, construction, and quality contro l. Apollo redesign kept American astronauts grounded until September 1968.

After the fire, NASA could no longer count on a friendly reception on Capitol Hill. The fire, plus growing pressure on the federal budget, meant that all NASA programs were subjected to increased oversight. In March, Aviation Week & Space Technology reported a "growing antipathy from Congress" toward NASA's programs, adding that "[d]elays in the manned program, resulting from the Apollo 204 crew loss . . . will hamper the agency's arguments before Congress since public interest will dwindle without spectacular results."[24] The magazine predicted, however, that Project Gemini's conclusion would free up funds in FY 1968, permitting "a modest start on Apollo Applications and . . . Voyager." [25]

As NASA in general came under increased scrutiny the piloted flyby concept suffered high-level criticism for the first time . The President's Science Advisory Committee (PSAC) report The Space Program in the Post-Apollo Period (February 1967) was generally positive, calling for continued Apollo missions to the Moon after the first piloted lunar landing , as well as planetary exploration using robots suc h as Voyager.[26] The PSAC reiterated Faget's 1962 criticism of the piloted flyby mission, however, stating that

> the manned Mars flyby proposal, among its other weaknesses, does not appear to utilize man in a unique role . . . it appears to us that NASA must address itself more fully to the question, "What is the optimum mix of manned and unmanned components for planetary exploration?"[27]

The PSAC also complained that Voyager and the Planetary JAG's piloted flyby plans were "distinct and apparently independent plans for planetary exploration," and criticized NASA for "absence of integrated planning in this area."[28] As has been seen, this criticism reached the Planetary JAG early enough for an integrated plan to be included in its report. NASA officials denied, however, that the PSAC's criticisms had prompted its effort to integrate Voyager and the piloted flyby.[29]

The PSAC's critique stung the Planetary J AG. One response was to distance itself from the term "flyby"—a word identified increasingly with automated explorers since Mariner 4's success—by dubbing its mission an "encounter."[30] Planetary JAG members also sought to reemphasize that the encounter mission astronauts would accomplish productive observations and experiments throughout their two-year voyage , not just during the hours of Mars encounter.

OMSF advanced planner Edward Gray and his deputy Franklin Dixon first publicly proposed the Planetary JAG's Apollo-based piloted Mars flyby as an FY 1969 new start the next month (Marc h 1967) at the AAS Fifth Goddard Memorial Symposium, where they presented a paper called "Manned Expeditions to Mars and Venus."[31] That same month, NASA forecast a stable annual budget of about $5 billion per year through 1970, after which the budget would dec line to $4.5 billion annually for the rest of the 1970s. New programs such as Voyager and the piloted flyby would be phased in as the share of NASA 's budget allotted to Apollo lunar missions decreased. [32] In May, Aviation Week & Space Technology reported that the $71.5-million new-start funding approved for Voyager by the House Space Committee "does not face serious problems."[33]

## A New Era for NASA

By the beginning of 1967, 25,000 United States service-men had died in Vietnam. The summer of 1967 sa w racial violence wrack Newark, New Jersey, and Detroit, Michigan. Large sections of Detroit burned to the ground. At least 5,000 people lost their homes , and more than 70 lost their lives. Violence also swept more than 100 other American cities. Detroit alone suffered up to $400 million in damage . Needless to sa y, most Americans focused more on Earth than on space .

The cost of the Vietnam War soared to $25 billion a year—the entire FY 1966 NASA budget every 10 weeks. This, plus the cost of President Johnson's Great Society social welfare programs , led to spiraling Federal budget deficits . Congress approached the Johnson Administration's 1968 Federal Budget with its scissors out, and NASA was an easy target.

In early July, Aviation Week & Space Technology reported that the H ouse and Senate had "sustained the pace of spending in the Apollo program but seri-ously cut into NASA 's plans for both manned and unmanned space programs of the future."[34] The Senate voted dow n all Voyager funding, while the House cut the program to $50 million . House and Senate conferees settled on $42 million for the auto-mated Mars program. In response, NASA announced that a 1971 Voyager mission was out of the question. A 1973 landing was, however, still feasible if the pro-gram was funded adequately in FY 1969.[35]

In early July, the Senate report on its FY 1968 NASA authorization bill specifically advised against piloted planetary missions, stating that "all near-term [piloted] missions should be limited to earth orbital activity or further lunar exploration."[36] Later that month, in testi-mony to the Senate Appropriations Committee, James Webb refused to "give aid and comfort to those who would cut our program" when asked by Spessard Holland (Democrat-Florida) to c hoose between $45

million for AAP and $50 million for Voyager. Holland chided Webb for "failing to see that Congress is faced with dilemmas in applying all its economies."[37]

That some in the aerospace world were sympathetic to Holland's plight is telling . In an editorial titled "New Era for NASA," for example, Aviation Week & Space Technology editor Robert Hotz wrote,

> We have no quarrel with reductions imposed so far by Congress . . . . They reflect a judicious and necessary pruning of NASA's budget . . . . [Space exploration] cannot hope to occupy such a large share of the national spotlight in the future as it did during the pioneering da ys of Mercury and Gemini when the w ar in Vietnam was only a tiny c loud on a distant horizon; when no American city cores had yet glowed red at night, and when a tax cut was the order of the day instead of the tidal w ave of tax rises that now threatens to engulf the nation.[38]

Though none of this augured well for piloted planetary missions, the Planetary J AG continued planning its piloted encounter mission with the aim of seeing it included as an FY 1969 new start. The revised Planetary JAG plan called for just two MS-IVBs.[39] This meant that only two Saturn V rockets would need to be launched in rapid succession, so the costly new Pad 39C was no longer required.

The encounter spacecraft would again inc lude an Experiment Module with an automated probe suite based on Voyager technology. This time, however, the probes, including at least one large MSSR lander , would be sealed in the Experiment Module before launch from Earth and sterilized to a void biological contamination of Mars . Previous piloted flyby studies had justified the presence of astronauts in part by their ability to service the probes during flight. This would now be impossible because servicing would introduce contamination.

The Planetary J AG realized that the MSSR w as the most challenging element of its encounter mission plan—the one demanding the earliest development start if the first piloted encounter mission w as to be ready for flight in 1975. On 3 August 1967, therefore, MSC issued a Request for Proposals for a "9-month engineering study . . . to perform a detailed analysis and preliminary design study of unmanned probes that would be launched from a manned spacecraft on a Mars encounter or a Mars capture mission,[and] would retrieve samples of the Mars surface and atmosphere and rendezvous with the manned spacecraft. " MSC added, "The results of this study" would "aid in selecting experiment payload combinations of these and other probes and in configuring the Experiment Module section of the manned spacecraft used in the Mars . . . Reconnaissance/Retrieval missions in the 1975-1982 time period." Cost and technical proposals were to be submitted to MSC by 4 September.[40] At the same time, MSC released an RFP calling for a piloted flyby spacecraft design study.

The Planetary JAG knew of the de facto congressional "no new starts" injunction but apparently assumed that it did not apply to studies with implications beyond the next fiscal year.[41] Congressman Joseph Karth (Democrat-Minnesota), chair of the House Subcommittee on Space Science and Applications, saw it differently. Normally a strong NASA supporter, he lashed out at the "ostrich-like, head-in-the-sand approach of some NASA planning," and added, "Very bluntly, a manned mission to Mars or Venus by 1975 or 1977 is now and always has been out of the question—and anyone who persists in this kind of misallocation of resources is going to be stopped."[42]

By August, the expected 1967 Federal budget deficit was $30 billion. Goaded by MSC's Request for Proposals, on 16 August 1967 the House zeroed out funding for both Voyager and advanced piloted mission planning. AAP funding fell to $122 million. On 22 August, the House approved a $4.59 billion FY 1968 NASA budget—a cut of more than $500 million from the January 1967 White House request.

Faced by a spiraling budget deficit, war and anti-war dissent, and urban riots, President Johnson reduced his support for NASA, saying, "Under other circumstances I would have opposed such a cut. However, conditions have greatly changed since I submitted my January budget request."[43] He added, "Some hard choices must be made between the necessary and the desirable . . . . We . . . dare not eliminate the necessary. Our task is to pare the desirable."[44]

## Denouement

The Voyager program died in part because NASA cast it as a lead-in to piloted flybys. The scientific community viewed Voyager's loss as a slap in the face. In September, in an unusual move, NASA officials went before the Senate Appropriations Committee to negotiate a Mariner mission in 1971 and a Mars landing mission in 1973, both designed "to conform to sharply reduced funding in FY 1969."[45] The 1971 Mariner mission became Mariner 9. In March 1968, NASA unveiled Project Viking—a cut-price version of the Voyager program. Viking, managed by NASA's Langley Research Center, emerged as one of the few FY 1969 NASA new starts.

MSC received and reviewed MSSR study proposals from industry, although, of course, no contract for such a study was ever issued. The piloted flyby mission, the object of so much study from mid-1962 to late 1967, was defunct. Despite the obvious congressional hostility toward advanced planning, however, NASA's piloted Mars mission studies were not. As will be seen in the next chapter, the focus shifted to the other area of Planetary JAG emphasis—piloted Mars landing missions using nuclear rockets.

# Chapter 5: Apogee

Thus, Mr. Vice President—[the s]olar system is opening up before us . With landing on the Moon we know that man can la y claim to the planets for his use. We know further that man will do this . The question is when? We know that [the] U.S. will take part. The question is how soon will we follow up on what we ha ve begun with Apollo? It could be the early 1980s. (Thomas Paine, 1969)[1]

## The Big Shot

In a November 1965 article on the next 20 years of space flight, Wernher von Braun sought to convey the Saturn V rocket's immense potential. "One Saturn V alone," he wrote, "will carry twice as much payload as the entire NASA space program up to this point in time. In fact, all the orbiters, all the deep space probes, and all the Mercurys and Geminis that have ever flown would only load the cargo compartment of one Saturn V to 50% of capacity."[2] With Saturn V available, the Moon, Mars, and indeed the entire solar system seemed within reach.

The first of fifteen Saturn V's ordered by NASA to support Project Apollo rolled out to Launc h Pad 39A at Kennedy Space Center on 26 August 1967. Designated AS-501, the mighty roc ket would launc h Apollo 4, the first unmanned test of an Apollo CSM spacecraft. The 24-hour countdown commenced early on 8 November and reached T-0 at 7 a.m. Eastern Standard Time on 9 November. Seen from the KSC press site , three and one-half miles from the pad, the white and blac k rocket rose slowly at the summit of an expanding mountain of red flame and gray smoke. Thunder from "the Big Shot," as the news media nic knamed AS-501, drowned out television and radio reporters giving live commentary and threatened to collapse their temporary studios.

AS-501 stood 111 meters tall and weighed about 2,830 metric tons at liftoff. Its 10-meter-diameter S-IC first stage carried 2,090 metric tons of kerosene fuel and liquid oxygen oxidizer for its five F-1 roc ket engines. They gulped 13.6 metric tons of propellants eac h second to develop a total of 3.4 million kilograms of thrust at liftoff. AS-501's first stage depleted its propellants in two and one-half minutes at an altitude of 56 kilometers, detached, and crashed into the Atlantic about 72 kilometers from Pad 39A.

The 10-meter-diameter S-II second stage carried 423 metric tons of liquid hydrogen and liquid oxygen for its five J-2 engines, which developed a total of 1 million pounds of thrust. The S-II depleted its propellants after six and one-half minutes at an altitude of 161 kilometers.

The 6.7-meter-diameter S-IVB third stage carried 105 metric tons of liquid hydrogen and liquid oxygen for its single restartable J-2 engine , which fired for two minutes to place the Apollo 4 CSM in a 185-kilometer parking orbit. For an Apollo lunar mission, the J-2 engine would ignite again after one orbit to place the Apollo spacecraft on course for the Moon. For Apollo 4, the third stage restarted after two Earth orbits, 3 hours and 11 minutes after liftoff, putting the stage and spacecraft into an Earth-intersecting ellipse with a 17,335-kilometer apogee (highest point above the Earth).

The Apollo 4 CSM separated from the S-IVB stage, then fired its engine for 16 seconds to nudge its apogee to 18,204 kilometers. The CSM engine ignited a second time 8 hours and 10 minutes into the flight to throw the CM at Earth's atmosphere at a lunar-return speed of about 40,000 kilometers per hour. The CM separated and positioned itself with its bowl-shaped heat shield forward. Heat shield temperature soared to 2,760 degrees Celsius, and CM deceleration reac hed eight times the pull of Earth' s gravity. Three parachutes opened, and the Apollo 4 CM splashed into the Pacific Ocean 10 kilometers from the planned spot, 8 hours and 38 minutes after liftoff.

The success of AS-501/Apollo 4 helped rebuild confidence in NASA's ability to fulfill K ennedy's mandate following the January 1967 fire. President Johnson told reporters that the "successful completion of toda y's flight has shown that we can launc h and bring bac k safely to Earth the space ship that will take men to the [M]oon." Von Braun told reporters that he regarded "this happy day as one of the three or four highlights of my professional life—to be surpassed only by the manned lunar landing."[3]

## "To the Very Ends of the Solar System"

Apollo 4 also cheered Mars planners, for Saturn V had become their launch vehicle of choice following the end of post-Saturn rocket planning in 1964. NASA and AEC engineers developing the NERV A nuclear-thermal

# Chapter 5: Apogee

rocket engine saw special cause for celebration, for Saturn V was their brainchild's ride into space. The encouragement was well timed. NERVA, which stood for Nuclear Engine for Rocket Vehicle Application, still had no approved mission and had just survived a narrow scrape in a Congress ill-disposed toward funding technology for future space missions.

NERVA was a solid-core nuclear-thermal rocket engine. Hydrogen propellant passed through and was heated by a uranium nuclear reactor, which caused the propellant to turn to plasma, expand rapidly, and vent out of a nozzle, producing thrust. Unlike chemical rockets, no oxygen was required to burn the hydrogen in the vacuum of space. Nuclear-thermal rockets promised greater efficiency than chemical rockets, meaning less propellant was required to do the same work as an equivalent chemical system. This would reduce spacecraft weight at Earth-orbit departure, opening the door to a broad range of advanced missions.

Initial theoretical work on nuclear-thermal rockets began at Los Alamos National Laboratory (LANL) in 1946. The New Mexico laboratory operated under the aegis of the AEC. The joint AEC-U.S. Air Force ROVER nuclear rocket program began in 1955, initially to investigate whether a nuclear rocket could provide propulsion for a massive intercontinental missile. In 1957, the solid-core reactor engine design was selected for ground testing. The test series engine was appropriately named Kiwi, for it was intended only for ground testing, not for flight.

Citing LANL's nuclear rocket work, AEC supporters in the U.S. Senate, led by New Mexico Democrat Clinton Anderson, pushed unsuccessfully in 1958 for the commission to be given control of the U.S. space program. Anderson was a close friend of Senate Majority Leader Lyndon Johnson, who led the Senate Space Committee formed after Sputnik 1's launch on 4 October 1957.[4] In October 1958, the Air Force transferred its ROVER responsibilities to the newly created NASA, and ROVER became a joint AEC-NASA program. AEC and NASA set up a joint Space Nuclear Propulsion Office (SNPO). NASA Lewis—which at this time was performing the first NASA Mars study, an examination of the weight-minimizing benefits of advanced propulsion, including nuclear rockets (see chapter 2)—became

responsible within NASA for technical direction of the ROVER program.

In July 1959, the first Kiwi-A test was carried out successfully using hydrogen gas as propellant at the Nuclear Rocket Development Station (NRDS) at Jackass Flats, Nevada, 90 miles from Las Vegas. Senator Anderson arranged for delegates to the Democratic National Convention to be on hand for the second Kiwi-A test in July 1960. At the Convention, Anderson arranged for a plank on nuclear rocket development to be inserted into the Democratic Party platform.[5] In October 1960, the third Kiwi-A test using hydrogen gas showed promising results, building support for a contract to be issued for development of a flight-worthy nuclear rocket engine.

The Democratic ticket of John Kennedy and Lyndon Johnson narrowly defeated Dwight Eisenhower's Vice President, Richard Nixon, in the November 1960 election. Anderson took over as head of the Senate Space Committee. President Kennedy embraced space after the Soviet Union helped end his White House honeymoon by launching the first human into space on 12 April 1961. He charged Johnson with formulating a visible, dramatic space goal the United States might reach before the Soviets. Johnson suggested landing an American on the Moon.

Before a special joint session of Congress on 25 May 1961, Kennedy called for an American astronaut on the Moon by the end of the 1960s. Then he asked for "an additional $23 million, together with $7 million already available, [to] accelerate development of the ROVER nuclear rocket. This gives promise of some day providing a means for even more exciting and ambitious exploration of space, perhaps beyond the Moon, perhaps to the very ends of the solar system . . . ."[6]

Because of Kennedy's speech, FY 1962 saw the real start of U.S. nuclear rocket funding. NASA and the AEC together were authorized to spend $77.8 million in FY 1962. Funding in the preceding 15 years had totaled about $155 million.

In July 1961, Aerojet-General Corporation won the contract to develop a 200,000-pound-thrust NERVA flight engine. NERVA Phase 1 occurred between July 1961 and January 1962, when a preliminary design was developed and a 22.5-foot NERVA engine mockup was assembled. At the same time, NASA Marshall set up the Nuclear Vehicle Projects Office to provide technical direction for the Reactor -In-Flight-Test (RIFT), a Saturn V-launched NERVA flight demonstration planned for 1967.

The first Kiwi-B nuc lear-thermal engine ground test using liquid hydrogen (December 1961) ended early after the engine began to blast sparkling, melting bits of uranium fuel rods from its reactor core out of its nozzle. Though the cause of this alarming failure remained unknown, Lockheed Missiles and Space Company w as made RIFT contractor in Ma y 1962. In early summer 1962 the Marshall Future Projects Office launc hed the EMPIRE study, motivated in part by a desire to develop missions suitable for nuc lear propulsion. Hence, early on NERVA became closely identified with Mars.

The second and third Kiwi-B ground tests (September 1962 and November 1962) failed in the same manner as the first. Failure cause remained uncertain, but vibration produced as the liquid hydrogen propellant flowed through the reactor fuel elements w as suspected.

The PSAC and the White House Budget Bureau allied against the nuclear rocket program following the third Kiwi-B failure. They opposed funding for an early RIFT flight test because they saw it as a foot in the door leading to a costly piloted Mars mission, and because they believed the technology to be insufficiently developed, something the Kiwi-B failures seemed to pr ove. Kennedy himself intervened in the AEC-NASA/Budget Bureau-PSAC deadlock, visiting Los Alamos and the NRDS in December 1962.

On 12 December 1962, Kennedy decided to postpone RIFT until after additional Kiwi-B ground tests had occurred, explaining that "the nuclear rocket . . . would be useful for further trips to the [M]oon or trips to Mars. But we have a good many areas competing for our available space dollars, and we have to channel it into those programs whic h will bring a result—first, our [M]oon landing, and then consider Mars ." Kennedy's decision marked the beginning of annual battles to secure continued nuclear rocket funding.[7]

At the Ma y 1963 AAS Mars symposium in Denver, SNPO director Harold F inger pessimistically reported that nuclear rockets were not likely to fly until the mid-1970s.[8] However, the fourth Kiwi-B test, in August 1963, revealed that vibration had indeed produced the earlier core failures. The problem had a relatively easy solution, so NASA, AEC, and nuclear engine supporters in Congress became emboldened. They pressed Kennedy to reverse his December 1962 decision.

William House, Aerojet-General's Vice President for Nuclear Rocket Engine Operations, felt sufficiently optimistic in October 1963 to tell the British Interplanetary Society's Symposium on Advanced Propulsion Systems that a Saturn V would launch a 33-foot-diameter RIFT test vehic le to orbit in 1967. He predicted that one NERV A stage would eventually be able to inject 15 tons on direct course to Mars, or 3 tons on a three-year flight to distant Pluto.[9]

Kennedy never had the opportunity to reconsider his RIFT decision. Following the young President' s November 1963 assassination, President Johnson took up the question. With an eye to containing government expenditures, he canceled RIFT in December 1963 and made NERVA a ground-based research and technology effort.

The year 1964 sa w the successful first ground test of the redesigned Kiwi-B engine and the first NERV A start-up tests. It also marked the nuc lear rocket program's peak funding year, with a joint AEC-NASA budget of $181.1 million. Though NERVA was grounded, work proceeded under the assumption that success would eventually lead to clearance for flight.

The nuclear rocket program budget gradually declined, dropping to $140.3 million in FY 1967. NERVA did not come under concerted attac k, however, until the bitter battle over the FY 1968 NASA budget. In August 1967, Congress deleted all advanced planning and Mars Voyager funds from NASA's FY 1968 budget because it saw them as lead-ins to a costly piloted Mars program, and Johnson refused to sa ve them (see c hapter 4). NERVA funding was eliminated at the same time.

Voyager had to wait until FY 1969 to be resurrected as Viking. Through Anderson's influence, however, NERVA did better—the nuclear rocket program was restored with a combined AEC-NASA budget of $127.2 million for FY 1968. As if to celebrate Anderson's intervention, the NRX-A6 ground test in December 1967 saw a NERVA engine operate for 60 minutes without a hitch.

## Boeing's Behemoth

In January 1968, the Boeing Company published the final report of a 14-month nuclear spacecraft study conducted under contract to NASA Langley. The study was the most detailed description of an interplanetary ship ever undertaken.[10] As shown by the EMPIRE studies, the propellant weight minimization promised by nuclear rockets tended to encourage big spacecraft designs. In fact, Boeing's 582-foot long Mars cruiser marked the apogee of Mars ship design grandiosity.

At Earth-orbital departure, Boeing's behemoth would include a 108-foot-long, 140.5-ton piloted spacecraft and a 474-foot-long propulsion section made up of five Primary Propulsion Modules (PPMs). The entire spacecraft would weigh between 1,000 and 2,000 tons, the exact weight being dependent upon the launch opportunity used. Each 33-foot-diameter, 158-foot-long PPM would hold 192.5 tons of liquid hydrogen. A 195,000-pound-thrust NERVA engine with an engine bell 13.5

feet in diameter would form the aft 40 feet of each PPM. The six-person piloted spacecraft would consist of a MEM lander, a four-deck Mission Module, and an Earth Entry Module.

Three PPMs would constitute Propulsion Module-1 (PM-1); two would constitute PM-2 and PM-3, respectively. PM-1 would push the ship out of Earth orbit toward Mars, then detach; PM-2 would slow the ship so that Mars' gravity could capture it into orbit, then it would detach; and PM-3 would push the ship out of Mars orbit toward Earth. At Earth, the crew would separate in the Apollo CM-based Earth Entry Module, reenter Earth's atmosphere, and splash down at sea.

Six uprated Saturn V rockets would place parts for Boeing's Mars ship in Earth orbit for assembly. Assembly crews and the flight crew would reach the spacecraft in Apollo CSMs launched on Saturn IB rockets. The 470-foot-tall uprated Saturn V, which would include four solid-fueled strap-on rockets, would

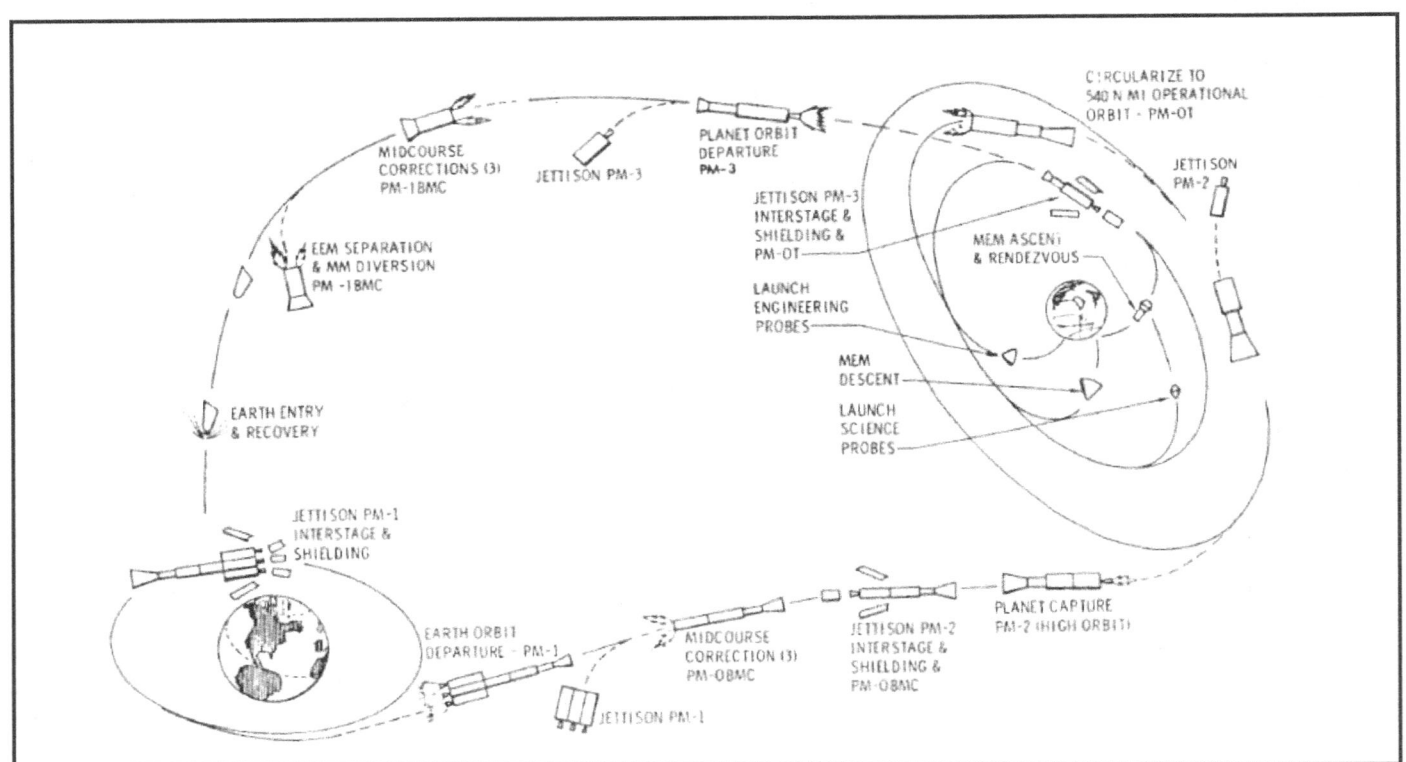

Figure 12—In January 1968, Boeing proposed this complex Mars expedition plan using nuclear rockets and an opposition-class trajectory. The company's Mars ship would measure nearly 200 meters long and support a crew of six. (Integrated Manned Interplanetary Spacecraft Concept Definition, Vol. 1, Summary, D2-113544-1, Boeing Company, Aerospace Group, Space Division, Seattle, Washington, p. 7.)

be capable of delivering 274 tons to a 262-mile circular Earth orbit. Boeing envisioned modifying KSC Saturn V launch pads 39A and 39B to launch the uprated Saturn V, and building a new Pad 39C north of the existing pads.

The company's report listed opportunities for nine Venus-swingby, one conjunction-class, and five opposition-class Mars expeditions between November 1978 and January 1998. The conjunction-class mission would last 900 days, while the Venus-swingby and opposition-class missions would last from 460 to 680 days.

Boeing envisioned using the MOR mission plan NASA Lewis used in its 1959-1961 studies. The MEM for descending to Mars from Boeing's orbiting Mars ship was designed for MSC between October 1966 and August 1967 by North American Rockwell (NAR), the

Apollo CSM prime contractor.[11] NAR's MEM report, published the same month as the Boeing report, was the first detailed MEM study to incorporate the Mariner 4 results. Cost minimization was a factor in NAR's MEM design. The company proposed a 30-foot-diameter lander shaped like the conical Apollo CM. The Apollo shape, it argued, was well understood and thus would require less costly development than a novel design.

The lightest NAR MEM (33 tons) would carry only enough life support consumables to support two people on Mars for four days, while the heaviest (54.5 tons) was a four-person, 30-day lander. Like the Apollo Lunar Module (and many previous MEM designs), NAR's MEM design included a descent stage and an ascent stage. The MEM would contain two habitable areas—the ascent capsule and the descent stage lab compart-

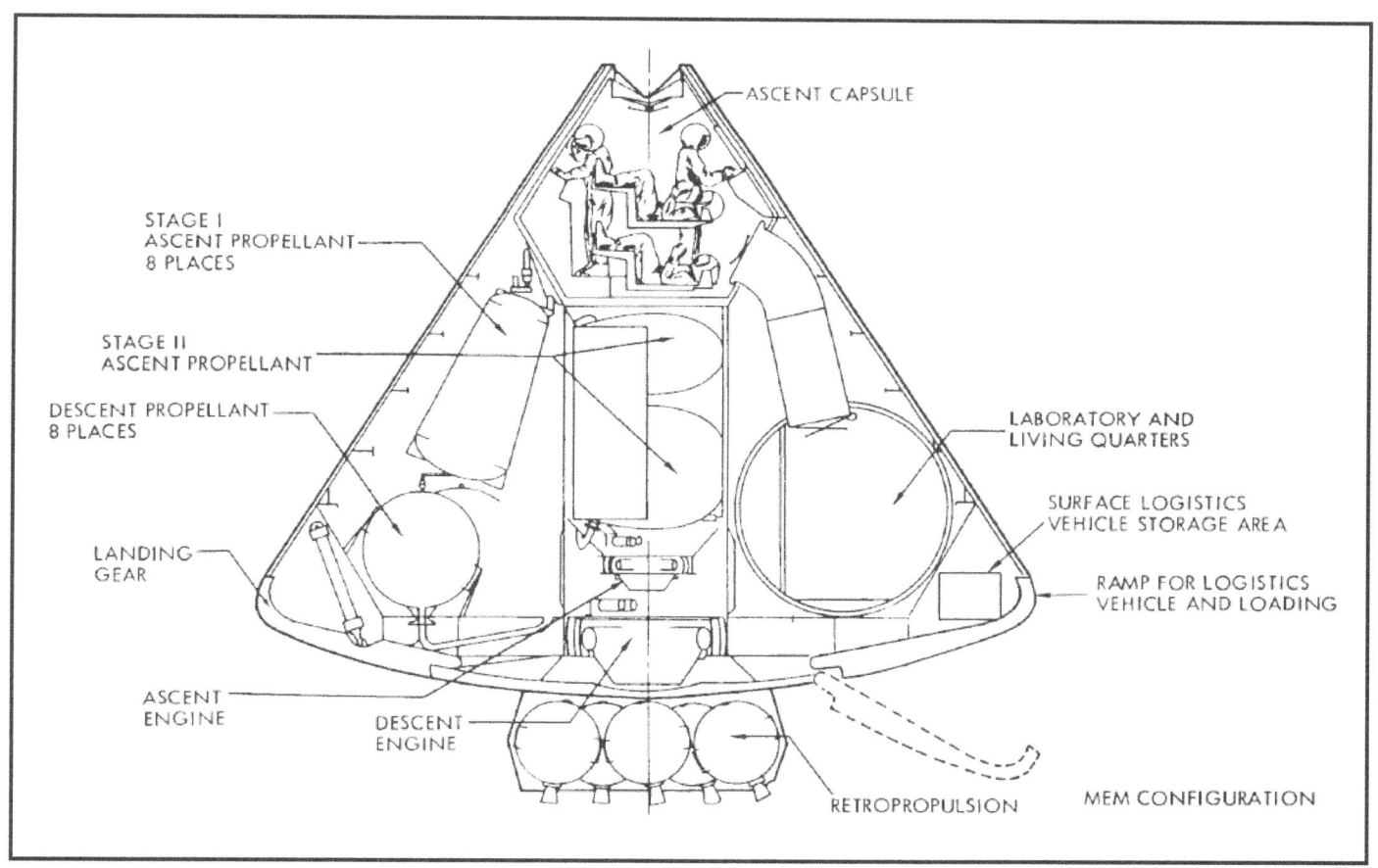

Figure 13—Cutaway of North American Rockwell's 1968 Mars lander. Based on the Apollo Command Module shape, its design incorporated new Mars atmosphere data gathered during the 1965 Mariner 4 automated Mars flyby. (Manned Exploration Requirements and Considerations, Advanced Studies Office, Engineering and Development Directorate, NASA Manned Spacecraft Center, Houston, Texas, February 1971, p. 5-3.)

ment. The ascent capsule would include an Apollo docking unit for linking the MEM to the mothership , and the lab compartment would inc lude an airloc k for reaching the Martian surface.

The MEM's Apollo-style bowl-shaped heat shield would protect it from friction heating during Mars atmosphere entry. To reduce cost, NAR proposed to develop a single heat shield design for both flight tests in Earth's atmosphere and Mars atmosphere entry. This meant, of course, that the shield would be more robust, and thus hea vier, than one designed

specifically for Mars atmosphere entry . During Mars atmosphere entry the crew would feel seven Earth gravities of deceleration.

After atmospheric entry, the MEM would slow its descent using a drogue parachute followed by a larger ballute (balloon-parachute). At an altitude of 10,000 feet the ballute would detac h. The MEM's descent engine would fire; then two of the astronauts would climb from their couches to stand at controls and pilot the MEM to touchdown. The company proposed using liquid methane/liquid oxygen propellants that would

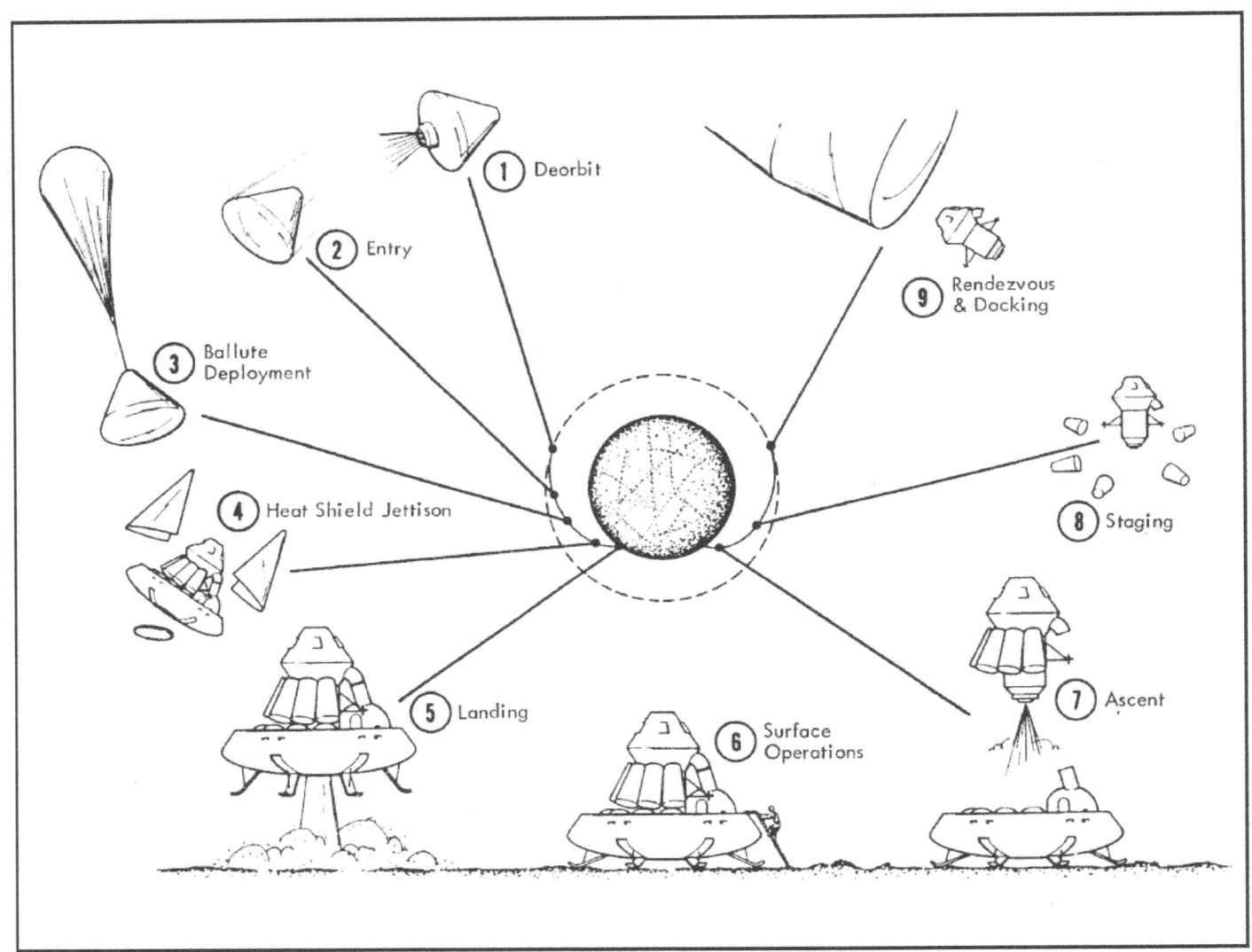

Figure 14—North American Rockwell's plan for landing on Mars and returning to Mars orbit. The company's lander, a two-stage design, would support up to f our astronauts on Mar s for up to 30 days and return to the orbiting mother ship with up to 300 pounds of rocks. (Integrated Manned Interplanetary Spacecraft Concept Definition, Vol. 4, System Definition, D2-113544-4, Boeing Company, Aerospace Group, Space Division, Seattle, Washington, January 1968, p. 145.)

offer high performance but not readily boil off or decompose. The MEM would carry enough propellants for two minutes of hover. Its six landing legs would enable it to set down safely on a 15-degree slope.

For return to the mothership in Mars orbit, the crew would strap into the ascent capsule with their Mars samples and data. The ascent stage engine would ignite, burning methane/oxygen propellants from eight strap-on tanks. The ascent stage would blast a way from the descent stage, climb vertically for five seconds, then pitch over to steer tow ard orbit. Once empty, the strap-on tanks would fall a way; the ascent engine would then draw on internal tanks to complete Mars orbit insertion and rendezvous and docking with the mothership.

NAR had MEM development commencing in 1971 to support a 1982 Mars landing. The company envisioned a MEM flight test program using six MEM test articles and a range of roc kets, including three two-stage Saturn Vs. The 1979 piloted MEM entry and landing test, for example, would have a fully configured MEM launched into Earth orbit on a two-stage Saturn V with a piloted CSM on top. In orbit the CSM would detac h, turn, and dock with the MEM for crew transfer. The crew would then cast off the CSM and fly the MEM to landing on Earth.

Boeing scheduled the first Mars expedition for 1985-1986, with Mars expedition contract a wards in 1976, and Mars hardware tests in low-Earth orbit beginning in 1978. NAR estimated development cost of its MEM at $4.1 billion, while Boeing's study placed total Mars program cost at $29 billion.

## End of an Era

As Aerospace Technology magazine put it in May 1968, "If the political c limate in Washington for manned planetary missions is as bleak as the initial congressional budget hearings indicate, the [NAR MEM] study is . . . likely to be the last of its type for at least a year."[12] In fact, it was the last until the late 1980s. As the battle over the FY 1968 budget during the summer of 1967 made abundantly clear, a $29-billion Mars program enjoyed support in neither the J ohnson White House nor the Congress. Events in 1968 made even

more remote the possibility that the US. might take on a new Apollo-scale space commitment.

On 30 J anuary 1968, immediately after Boeing and NAR published their reports, North Vietnam invaded South Vietnam on the eve of Tet, the lunar new year. Though repulsed by U. S. and South Vietnamese forces, the large-scale offensive drove home to Americans and the Johnson White House that American involvement in Indochina would likely grow before it shrank.

At the end of May, the Defense Department asked for a $3.9-billion supplemental appropriation. Of this, $2.9 billion w as earmarked to pa y for the Tet Offensive—the Defense Department needed, for example, to replace 700 destroyed helicopters—while $1 billion would beef up U.S. defenses in South Korea following the Pueblo incident, in which North Korea seized a U.S. ship.[13] A total of 14,592 American soldiers had been killed in Vietnam by the close of 1968, by which time the total U.S. forces in Indochina stood at more than half a million.

There was also trouble at home. Johnson was a political casualty of Tet and other troubles shaking the nation. On 31 Marc h 1968, he announced that he would not stand for reelection. On 4 April 1968, civil rights leader Martin Luther King Jr. was gunned down in Memphis, Tennessee; his death triggered racial violence across the country. That same month students at Columbia University in New York seized buildings to protest the Vietnam War in one of more than 200 major demonstrations at some 100 universities during the year. On 6 June 1968, Democratic Party front-runner Robert K ennedy was shot in Los Angeles. In August, antiwar protesters disrupted the Democratic National Convention.

Near the start of the FY 1969 budget cyc le in early February 1968, as American and South Vietnamese forces pushed back the North Vietnamese, James Webb testified to the House Space Committee, where a $4 billion FY 1969 NASA budget w as, according to one committee staffer, a "fait accompli." He reminded the Committee that

> NASA's 1969 authorization request, at the $4.37-billion level, is $700 million below the amount requested last year. NASA expenditures for Fiscal 1969 will be down $230 million

from this year, $850 million from last year, and $1.3 billion less than in Fiscal 1966. The NASA program has been cut. I hope you will decide it has been cut enough . . . .[14]

In testimony to the Senate Appropriations Committee in May, after the House approved a $4-billion NASA budget, Webb told the Senators that President Johnson had directed him to acq uiesce to the cut, then expressed concern over NERVA's future.[15] The nuclear rocket stayed alive in early J une 1968 only after a lengthy Senate floor battle w aged by Howard Cannon (Democrat-Nevada), whose state inc luded the NRDS. Webb told the Senate Appropriations Committee later that month that the $4-billion NASA budget would require halting Saturn V production for a year and canceling NERVA. In an attempt to rally NERVA supporters to approve their engine's ride into space, he added that "to proceed with NERV A while terminating Saturn V cannot be justified."[16]

On 1 August 1968, Webb turned down George Mueller's request to make long lead-time purchases for manufacture of two more Saturn V's, the sixteenth and seventeenth in the series. He informed the OMSF c hief that production would halt with the fifteen already allotted for the Apollo lunar program.[17] A week later Webb told Congress that "the future is not bright" for the Saturn rockets.[18]

At a White House press conference on 16 September 1968, Webb announced that he would step down after nearly 8 years as NASA Administrator. He told journalists that he left the Agency "well prepared . . . to carry out the missions that have been approved . . . . What we have not been able to do under the pressures on the budget has been to fund new missions for the 1970s . . . ."[19]

## Thomas Paine Takes Charge

The final FY 1969 NASA budget w as $3.995 billion, making it the first below $4 billion since 1963.This was more than $370 million below NASA 's request, but almost exactly what J ohnson had told Webb to accept in May. The Saturn V production line went on standby. The nuclear rocket program received $91.1 million, of which $33.1 million came from NASA funds.

NASA Deputy Administrator Thomas Paine became Acting NASA Administrator upon Webb's departure on 7 October. Webb, a 25-year veteran of F ederal government service, had described P aine as one of a "new breed of scientist-administrators making their way into government."[20] Formerly director of General Electric' s TEMPO think tank, he had entered government service through a program for recruiting managers from industry. Paine had become Webb's Deputy Administrator in March 1968, replacing Robert Seamans. When he took over NASA from Webb, Paine had seven months of Federal government experience.

Immediately after taking NASA's reins, Paine told the Senate Space Committee that he would seek a $4.5 billion NASA budget in FY 1970, followed by annual increases leading to a $5.5-billion budget in FY 1975. Paine said that he wanted a six- to nine-man space station serviced by Apollo CSMs in the mid-1970s. George Mueller also testified, calling for a $4.5-billion NASA budget in FY 1970. He said that this w as necessary to avoid a gap in piloted flights after the Apollo lunar landings.[21]

On 30 October 1968, the Budget Bureau completed a "highlights" paper on "major aspects of National Aeronautics and Space operations whic h warrant attention at an early point in 1969" for President Johnson's successor. The paper noted that "pressure is mounting to budget significant sums for follow-on manned space flight activities." It stated that "the advantages of nuc lear propulsion do not begin to approximate the costs for missions short of a manned Mars landing. No national commitment has been made to undertake this mission[,] whic h would cost $40-$100B[illion] . . . nevertheless, pressures are strong in NASA, industry, and Congress to undertake the development of the nuclear rocket."[22]

Republican Richard Nixon defeated Hubert Humphrey, Johnson's Vice President, for the White House in November. Though Apollo 7 had triumphantly returned NASA astronauts to orbit in October , space had been overshadowed as a campaign issue by the war, the economy, student revolt, and many other "down-to-Earth" issues. Nixon had promised a tax cut, which promised to place yet more pressure on F ederal agencies to cut spending.

Six weeks after the electi on, in the J ohnson Administration's twilight da ys, space flight won bac k the front page. On 21 December 1968, Apollo 8 astronauts Frank Borman, James Lovell, and William Anders became the first people to launc h into space on a Saturn V rocket and the first humans to orbit a world other than Earth. The Apollo 8 CSM dropped behind the Moon early on 24 December and fired its engine for four minutes to slow down and allow the Moon' s gravity to capture it into lunar orbit.

Thirty-five minutes after the spacecraft passed beyond the Moon's limb, it emerged from the other side . As it did, Earth rose into view over the hilly lunar horizon, and the crew snapped their planet' s picture. Lovell described the Moon to people on Earth as "essentially gray, no color; looks like plaster of Paris or sort of grayish deep sand."[23] Later, in one of the most memorable moments of the space age, the crew took turns reading to the world from the biblical book of Genesis. Early on Christmas Day 1968, after 10 lunar orbits, the Apollo 8 crew fired their CSM' s engine to escape the Moon' s gravitational pull and fall back to Earth.

Originally Apollo 8 was intended as an Earth-orbital test of the Saturn V and the Lunar Module Moon lander, but the Lunar Module w as not ready. Sending Apollo 8 to orbit the Moon w as first proposed in August 1968 by George Low, director of the Apollo Spacecraft Program Office at MSC, and was eagerly promoted by Tom Paine despite initial skepticism from NASA Administrator Webb.[24] Because the crew lacked a Lunar Module , they lacked the backup propulsion and life support systems it could provide. These would come in handy during James Lovell's next flight to the Moon on Apollo 13 in April 1970.

The image of Earth rising into view over the pitted gray Moon featured prominently on end-of-year magazines and newspapers. It formed a counterpoint of fragile beauty and bold human ac hievement that accentuated the war, dissent, and assassinations of 1968. This was reflected in Nixon' s first inaugural speec h on 20 January 1969:

> We have found ourselves rich in goods, but ragged in spirit; reaching with magnificent precision for the Moon, but falling into r aucous discord on Earth. We are caught in war, wanting peace. We are tor n by divisions, wanting unity.[25]

Democrat Paine submitted his resignation pro forma when Republican Nixon took office. Surprisingly, Nixon did not accept it. Though Aviation Week & Space Technology reported that Nixon w as impressed by the job Paine had done since coming to NASA, the real reasons were apparently less meritorious .[26] Nixon had never shown much interest in space and could find no ideologically suitable replacement who w anted to head NASA. He may also have desired to have a Democrat in place to blame if the Kennedy/Johnson Apollo program failed.[27] Paine was confirmed as NASA Administrator in March 1969.

Being a Democrat in a Republican administration was enough to lea ve Paine in a weak position. On top of that, however, Paine was a Washington neophyte. Webb had been wily , a Washington insider given to deal-making; Paine was an idealist given to emotive arguments. Paine was, according to NASA Historian Roger Launius, "every bit as zealous for his cause as had been his namesake." Furthermore, he was "unwilling to compromise and . . . publicly critical of the [Nixon] administration's lack of strong action" with regards to space.[28] He excoriated his Center directors for lacking boldness. He considered this disloyal to his view of America, the expansive country, ready to tackle any challenge.[29]

To Paine, the late 1960s was not a time to try men's souls. He complained to the Washington Evening Star of "what I would call almost a national hypoc hondria . . . in many ways crippling some of the forward-looking things we're able to do . . . I feel that one of the very highest priority matters is the w ar on poverty and the problems of the cities. But in the meantime we're making . . . a lot of progress in the civil rights area and really , this nation is a good deal healthier than we're giving it credit for today."[30]

Paine tried to use the excitement generated by Apollo 8 as a lever to gain Nixon' s commitment to an expansive post-Apollo future for NASA. His efforts were countered by voices counseling caution. Nixon had appointed "transition committees" to help chart a course for his new Administration. On 8 January 1969, the Task Force on Space transition committee, chaired by Charles Townes, handed in its report. The Task

Force, made up of 13 technologists and scientists, recommended against new starts and proposed a steady NASA budget of $4 billion per year "a rather frugal amount" equivalent to "three-quarters of one percent of GNP [Gross National Product]."[31]

The Task Force counseled continued lunar exploration after the initial Apollo Moon landings and advised Nixon to postpone a decision on a large space station and the reusable shuttle vehic le needed to resupply it economically. The primary purpose of the station was, it said, "to test man's ability for an extended spaceflight over times of a year or more, so that the practicality of a manned planetary mission could be examined. However, the desirability of suc h a mission is not yet clear . . . ."[32]

The Task Force recommendations resembled those in the February 1967 PSAC report, and with good reason—the membership lists of the two groups were almost identical. One new addition w as Robert Seamans, Secretary of the Air Force, who had been NASA Deputy Administrator when the PSAC had submitted its 1967 report.

Even as the Task Force presented its recommendations to Nixon, Paine's optimistic plans for NASA's FY 1970 budget foundered. President Johnson's FY 1970 budget request for NASA, released 15 January 1969, was $3.88 billion—$800 million less than the $4.7 billion "optimum" figure Paine had given the Budget Bureau in November and more than $100 million less than what Paine had said w as the "minimum acceptable." When Nixon's Budget Bureau chief, Robert Mayo, asked agency heads a week later to further trim the J ohnson budget, Paine pushed for a $198-million increase. Mayo quickly rebuffed P aine's request.[33] Nixon's FY 1970 budget went to Congress on 15 April. NASA's share was $3.82 billion, of which Congress eventually appropriated $3.75 billion.

## Space Task Group

Paine pointed to the Task Force on Space report as an example of what he did not want for NASA's future.[34] At a NASA meeting on space stations held in February at Langley, Paine invoked instead von Braun' s Collier's articles.[35] Following the meeting , Aviation Week & Space Technology magazine reported that NASA planned a 100-person space station by 1980, with first 12-person module to be launched on a modified Saturn V in 1975.[36]

Nixon's science advisor, Lee Dubridge, tried to get authority to set NASA's future course, in part because he sensed Paine's aims were too expansive, but Paine protested. On 13 February 1969, President Nixon sent a memorandum to Dubridge, Paine, Defense Secretary Melvin Laird, and Vice President Spiro Agnew, asking them to set up a Space Task Group (STG) to provide advice on NASA's future.[37] On 17 February, Nixon solicited Paine's advice on the agency's direction. Paine's long, detailed letter of 26 F ebruary sought to step around the STG process and secure from Nixon early endorsement of a space station. [38] In his response, Nixon politely reminded Paine of the newly formed STG.[39]

STG meetings began on 7 March 1969. In addition to the four voting members , the group inc luded observers: Glenn Seaborg of the AEC; U. Alexis Johnson, Under Secretary of State for Political Affairs; and, most influential, the Budget Bureau' s Mayo. Robert Seamans stood in for Melvin Laird. The STG chair was Agnew, another Washington neophyte. Misreading the Vice President's importance within the Nixon Administration, Paine focused his efforts on wooing Agnew to his cause . Much of the STG' s work was conducted outside formal STG meetings , which occurred infrequently.

NASA's STG position became based on the Integrated Program Plan (IPP) developed by Mueller' s OMSF, which was first formally described to Paine in a report dated 12 May.[40] Mueller attributed many of its concepts to a NASA Science and Technical Advisory Council meeting held in La Jolla, California, in December 1968. Though concerned mostly with Earth-orbital and cislunar missions, the report proposed that "the subsystems, procedures and even vehic les" for such missions "be developed with a view towards their possible use in a future planetary program . . . ."[41]

The IPP schedule was aggressive even by 1960s Moon race standards. Between 1970 and 1975, NASA would conduct a dozen Apollo lunar expeditions and launc h and operate three AAP space stations—two in Earth orbit and one in lunar polar orbit. The year 1975 would see the debut of the reusable Earth-orbital Space

Shuttle, which could carry a 25-ton, 40-foot-long, 22-foot-wide payload in its cargo bay.

Shuttle payloads would include a standardized space station module housing up to 12 astronauts, a propulsion module usable as a piloted Moon lander or Space Tug, and tanks containing liquid hydrogen propellant for the NERVA-equipped Nuclear Shuttle, which would first reach Earth orbit on an uprated Saturn V in 1977. Significantly, Mueller's IPP gave NERVA a non-Mars mission as part of a larger reusable transportation system in cislunar space. Up to 12 astronauts would conduct a Mars flight simulation aboard the Space Station in Earth orbit from 1975 to 1978, and 1978 would see establishment of a Lunar Base.

By 1980, 30 astronauts would live and work in cislunar space at any one time. Four Nuclear Shuttle flights and 42 Space Shuttle flights per year would support the Space Station Program. Six Nuclear Shuttle flights, 48 Space Shuttle flights, and eight Space Tug Moon lander flights per year would support the Lunar Base Program.

## NASA's Big Gun

Paine liked Mueller's ambitious IPP. He asked Wernher von Braun to make it even more expansive by building a Mars mission concept onto it in time for a 4 August presentation to the STG. The presentation was timed to capitalize on the enthusiasm and excitement generated by the first Apollo Moon landing mission, which was set to lift off on 16 July 1969.

Paine saw von Braun as "NASA's big gun." He believed that the space flight salesmanship for which the German-born rocketeer was famous could still help shape the future of American space flight as it had in the previous two decades. According to Von Braun, "it was an effort of a very few weeks to put a very consistent and good and plausible story together."[42]

Meanwhile, Paine's efforts to woo Agnew were, it appeared, beginning to pay off. At the Apollo 11 launch, the Vice President spoke of his "individual feeling" that the United States should set "the simple, ambitious, optimistic goal of a manned flight to Mars by the end of the century."[43]

On 20 July, Apollo 11 Commander Neil Armstrong and Lunar Module Pilot Edwin "Buzz" Aldrin landed the spider-like Lunar Module Eagle on the Moon's Sea of Tranquillity. At the start of humanity's first two-hour Moon walk, Aldrin described the landscape as a "magnificent desolation." The astronauts remained at Tranquillity Base for 21 hours before rejoining Command Module Pilot Michael Collins aboard the CSM Columbia in lunar orbit. On 24 July 1969, they splashed down safely in the Pacific Ocean, achieving the goal Kennedy had set eight years before.

In reporting the Apollo 11 landing, the Los Angeles Herald-Examiner pointed to space-age spin-offs, such as "new paints and plastics," then predicted that "the Mars goal should bring benefits to all mankind even greater than the . . . [M]oon program."[44] The Philadelphia Inquirer anticipated opposition to a Mars program; it asked, "will the inspiration be abandoned before the veiled censure of those who seem to suggest the solution of all human dilemmas lies in turning away from space to other priorities?"[45]

Aviation Week & Space Technology reported that "[s]pace officials sense that public interest is near an all-time high . . . ."[46] Yet polls taken at the time did not indicate strong public support for Mars exploration. A Gallup poll showed that the majority of people polled aged under 30 years favored going on to Mars; however, a larger majority of those over 30 opposed. Taken together, 53 percent of Americans opposed a Mars mission, 39 percent favored it, and 8 percent had no opinion.[47]

In addition to the polls, new automated probe data supplied Mars mission detractors with ammunition. The Mariner 6 spacecraft had left Earth on 24 February, just before STG meetings began. On 31 July 1969, as Paine and von Braun put the finishing touches on their 4 August pitch, it flew over the southern hemisphere of Mars, snapping 74 grainy images of a forbidding landscape pocked by craters. A feature known to Earth-based telescopic observers as Nix Olympica ("the Olympian Snows") appeared as a 300-mile crater with a bright central patch.

The spacecraft's twin, Mariner 7, had left Earth on 26 March. It flew over Mars' southern hemisphere on 5 August 1969, snapping 126 images of the smooth-floored Hellas basin, the heavily cratered Hellespontus region, and the south pole ice cap. The probes seemed

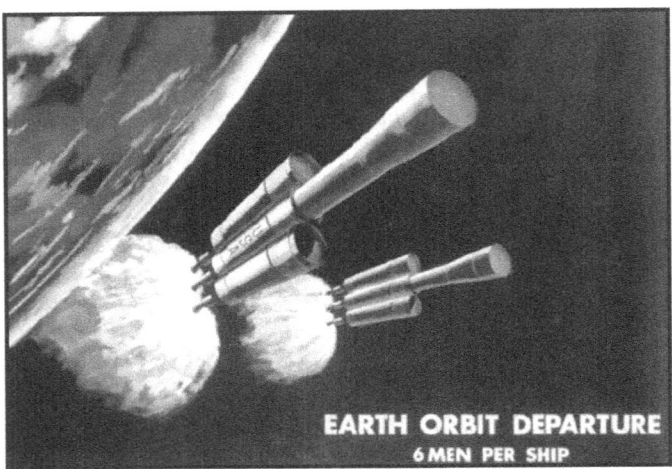

Figure 15—Twin Mars ships blast their all-male crews from Earth orbit using NERVA nuclear rocket stages. In August 1969, Wernher von Braun used images such as this to present NASA's vision of a Mars expedition in the 1980s to the Space Task Group and to Congress. (NASA Photo MSFC-69-PD-SA-176)

to confirm the pessimistic picture painted by Mariner 4 in 1965. The New York Times noted that NASA had "begun drumming up pressure to spend huge sums required to send men to Mars in the early 1980s. . . . But the latest Mariner information makes the possibility of life on Mars much less than it seemed even a week ago, thus removing much of the original motivation for such a project."[48]

NASA's 4 August STG presentation had three parts, lasted 55 minutes, and took into account neither the opinion polls nor the new Mars data. In the first part, Paine spent 20 minutes describing the "mystery, challenge, rich potential, and importance to man of the solar system" and "how the United States can move from [the] start represented by Apollo to exploration of the entire solar system with a program requiring only a modest investment of our national resources."[49]

Von Braun followed Paine and spent 30 minutes describing a piloted Mars expedition in 1982. His presentation formed the heart and soul of NASA's STG pitch.[50] In retrospect, it also marked the apogee of von Braun's career.

Von Braun drew on a sizable library of conceptual Mars spacecraft art generated in the Marshall Future Projects Office to show Mayo, Dubridge, Seamans, Johnson, Seaborg, and Agnew vehicles similar to the

Boeing Mars cruiser and the NAR MEM. In his IPP-based plan, the MEM was the only piece of hardware applicable only to Mars flight. All other vehicle elements would, he explained, be developed for cislunar roles. MEM go-ahead in 1974 would mark de facto commitment to a 1982 Mars expedition. The first space station module, the design of which would provide the basis for the Mars ship Mission Module, would fly in 1975, as would the first Earth-orbital Space Shuttle. The year 1978 would see the MEM test flight; then, in 1981, the first Mars mission would depart Earth orbit for a Mars landing in 1982.

The Mars mission would employ two Mars spacecraft consisting of three Nuclear Shuttles arranged side by side and a Mission Module. The complete spacecraft would measure 100 feet across the Nuclear Shuttles and 270 feet long. All modules would reach orbit on upgraded Saturn V rockets. After the twin expedition ships were assembled, reusable Space Shuttles would launch water, food, some propellant, and two six-person crews to the waiting Mars ships. At Earth-orbit launch,

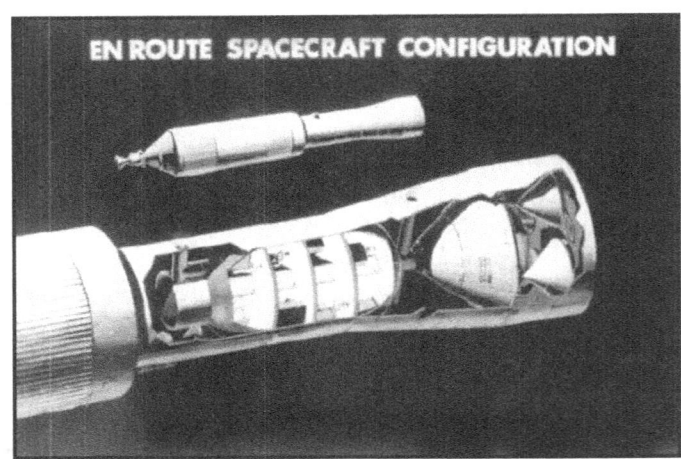

Figure 16—Compared with cramped Apollo spacecraft, the lodgings proposed for NASA's 1980s Mars ships were palatial. In this cutaway, note the four-deck Mission Module (center) and large conical Mars lander (right). (NASA Photo S-69-56295)

each ship would mass 800 tons, of which 75 percent was hydrogen propellant.

Von Braun targeted Mars expedition departure for 12 November 1981. The port and starboard Nuclear Shuttles would then fire their NERVA engines, achieve

Trans-Mars Injection, and shut down and separate from the center Nuclear Shuttle and Mission Module. They would turn around and fire their engines again to slow down and enter elliptical Earth orbit. A few days later they would reach perigee (lowest point above the Earth) at the original assembly orbit altitude, fire their engines to circularize their orbits, and rendezvous with the Space Station for refurbishment and re-use. The ships would weigh 337.5 tons each after port and starboard Nuclear Shuttle separation.

As in the Planetary JAG piloted flyby missions, the nine-month coast to Mars would be "by no means an idle phase." The ships each would serve as "a manned laboratory in space, free of the disturbing influences of the Earth." According to von Braun, "[t]he fact that there will be two observation points, Earth and spacecraft, permits several possible experiments." In addition, "as yet unidentified comets might be observed for the first time."[51]

Von Braun had the twin Mars ships reaching Mars on 9 August 1982. Each would fire the NERVA engine on its remaining Nuclear Shuttle to slow down and enter Mars orbit. At Mars Orbit Insertion each spacecraft would weigh 325 tons. The crews would then spend two days selecting landing sites for the expedition's 12 automated Sample Return Probes. The probes would land, retrieve samples uncontaminated by human contact, and lift off, then deliver the samples automatically to sterilized bio-labs on the ships for study.

If the samples contained no hazards, a three-man landing party would descend to the surface in one of the 47.5-ton MEMs. The other would be held in reserve—von Braun explained that "capability is provided for one man to land a MEM and bring a stranded crew back to the ship." He promised that "Man's first step on Mars will be no less exciting than Neil Armstrong's first step on the Moon."[52]

The astronauts would then spend between 30 and 60 days on Mars. Von Braun listed objectives for Martian exploration, including the following:

- Understand Martian geology "because Mars probably closely paralleled the earth in origin and . . . development."

- Search for life—von Braun stated that "preliminary data indicate that some lower forms of life

can survive in the Martian environment . . . in isolated areas higher forms . . . may exist. Man on Mars will [also] be able to study . . . the behavior of terrestrial life forms transplanted to the Martian environment."

- "Drilling for . . . water will be an early objective . . . and its discovery would open many possibilities . . . . For example, it might become possible to produce rocket fuel for the return trip on later missions."[53]

The landing party would lift off in the MEM ascent stage using the descent stage as a launch pad. The ascent stage would dock with the orbiting ship and the crew would transfer 900 pounds of samples and equipment, then would discard the expended ascent stage. The ships would ignite their center Nuclear Shuttles to leave Mars on 28 October 1982, after 80 days near the planet. The ships would weigh 190 tons each prior to Mars orbit departure.

Von Braun told the STG that the twin Mars ships would fly by Venus on 28 February 1983, to use the planet's gravity to slow their approach to Earth, thereby reducing the amount of braking propellant needed to enter Earth orbit. During swingby the astronauts would map Venus' cloud-shrouded surface with radar and deploy four automated probes.

Von Braun scheduled return to Earth for 14 August 1983. He noted that an Apollo-style direct reentry was possible; however, until "a better assessment can be made of the back contamination hazard (the return by man of pathogens that might prove harmful to earth inhabitants), a more conservative approach has been planned, i.e., the return of the crew to earth orbit for a quarantine period."[54] The center Nuclear Shuttles would place the Mission Modules in Earth orbit and perform rendezvous with the Space Station, where doctors would examine the astronauts. The Mars ships would weigh 80 tons each at mission's end, one-tenth of their Earth-departure weight. Following their quarantine period, the crew would return to Earth aboard a Space Shuttle. The center Nuclear Shuttles, meanwhile, would be refurbished and reused.

He then looked beyond the first expedition, stating that additional flights to Mars could occur during the periods 1983-84, 1986-87, and 1988-89. The 50-person Mars Base might be established in 1989, in time for the 20th anniversary of von Braun's presentation. Von Braun told the STG that NASA's budget would peak at $7 billion per year in 1975, or about 0.6 percent of GNP, and that it would level out at $5 billion in 1989, at which time its share of GNP would be 0.3 percent.[55] This assumed steady 4 percent annual growth in the U.S. economy. In his closing remarks, Paine put the cost a little higher than had von Braun; he told the other STG members that "[t]his kind of program would be possible for the United States with a budget rising to about $9 billion [per year] in the last half of the decade."[56]

## "Now Is Not the Time . . ."

NASA's vision was breathtaking, but stood little chance of acceptance in 1969 America. Robert Seamans appears to have been generally sympathetic to Paine's vision, yet cognizant of political and economic realities. He arrived at the 4 August meeting with a letter for Agnew laying out a less expansive view of America's future in space—one similar to the recommendations made by the transition Task Force in January. Seamans wrote, "I don't believe we should commit this Nation to a manned planetary mission, at least until the feasibility and need are more firmly established. Experience must be gained in an orbiting space station before manned planetary missions can be planned." Then he recommended against early commitment to a space station.

Seamans advised instead that NASA should expand AAP and continue lunar exploration "on a careful step-by-step basis reviewing scientific data from one flight before going to the next." He differed from the transition Task Force by recommending "a program to study by experimental means including orbital tests the possibility of a Space Transportation System that would permit the cost per pound in orbit to be reduced by a substantial factor (ten times or more)."[57] Aviation Week & Space Technology had by this time already predicted that the STG would recommend a reusable Space Shuttle as NASA's post-Apollo focus.[58]

On 5 August, the day Mariner 7 flew past Mars, Paine and von Braun presented their pitch to the Senate Space Committee. Clinton Anderson, its chair, had in effect already responded to the presentation; on 29 July 1969, he said that "now is not the time to commit ourselves to the goal of a manned mission to Mars."[59] Coming from Anderson, this was ominous and some-

what puzzling. The New Mexico Senator had backed NASA since its birth, in large part because the Agency gave the nuclear rocket program he supported funding and a raison d'être. His rejection of Mars placed him in a dilemma—how could he back nuclear propulsion yet not support what was widely seen as its chief mission? Other Space Committee members had similar reactions to NASA's presentation. Senator Mark Hatfield (Republican-Oregon) told Paine and von Braun that he supported the space program, but was "not really ready, at this point . . . to make commitments . . . to meet a deadline to get a man to Mars." Senator Margaret Chase Smith (Republican-Maine) named Paine's game, saying that the government "should avoid making long-range plans during this emotional period [following Apollo 11] . . . otherwise we may become involved in a crash program without the justification we had for Apollo—and therefore without the full support of Congress."[60]

Despite the clear signals from Congress, the STG remained split between Washington neophytes and old hands, with the former stubbornly preaching Mars and the latter counseling something less expansive. Robert Mayo broke the deadlock when he proposed that the group offer the President several pacing options contingent on available funds.[61]

Paine and Mueller then took their case to the public with a presentation to the National Press Club. Mueller painted a picture of NASA's space activities in 1979, when, he said, more than 200 people would work in space at one time. Most would be scattered in facilities between Earth orbit and the lunar surface; however, 12 would be en route to Mars in two ships.[62] Aviation Week & Space Technology editor Robert Hotz attended the Press Club talks and became swept up in NASA's vision. In his editorial following the talks he took a page from Paine's book, writing that

> the Apollo 11 mission has opened an endless frontier which mankind must explore. Man is extending his domain from the 8,000-mile-diameter of his home planet earth to the 8-billion-mile diameter of the solar system . . . . Hopefully [the President] will note that only by setting extremely high goals have extraordinary results been achieved . . . . We think Dr. Paine made a telling point when he warned against establishing future goals too low.[63]

Congress, meanwhile, voiced more reservations. George Miller (Democrat-California), chair of the House Committee on Science and Astronautics, did not want "to commit to a specific time period for setting sail to Mars." Miller was not opposed to going to Mars on principle; in fact, he believed it "highly probable that five, perhaps 10 years from now we may decide that it would be in the national interest to begin a carefully planned program extending over several years to send men to Mars."[64]

J. W. Fulbright (Democrat-Arkansas), Committee on Foreign Relations chair, sought to put Apollo in proper perspective as an element of 1960s realpolitik: "The [Apollo 11] landing called forth a great deal of poetizing about the human spirit bursting earthly bounds . . . . In all this I perceive not humbug . . . but rather more sententiousness than plain hard truth. Americans went to the Moon for a number of reasons of which, I am convinced, the most important by far was to beat the Russians."[65] Sending American astronauts on to Mars had nothing to do with beating the Russians Therefore, Fulbright saw little cause to support such a mission.

## America's Next Decades in Space

NASA released its report America's Next Decades in Space: A Report to the Space Task Group on 15 September 1969.[66] Paine was the principal author of the report, which aimed to promote NASA's STG position. In retrospect the report marked the apogee of NASA Mars expedition planning. With a note of pride it pointed out that, in NASA's first decade,

> the American space program progressed from the 31-pound Explorer 1 in earth orbit to Apollo spacecraft weighing 50 tons sent out to the moon [and] from manned flights of a few thousand miles and 15-minute duration to the 500,000 mile round-trip 8-day [Apollo 11] mission which landed men on the [M]oon and returned them safely to [E]arth.[67]

The NASA report then appealed to President Nixon to think of his place in history, and to see his decision as an unprecedented opportunity:

> At the moment of its greatest triumph, the space program of the United States faces a crucial situation. Decisions made this year will

affect the course of space activity for decades to come . . . . This Administration has a unique opportunity to determine the long-term future of the Nation's space progress. We recommended that the United States adopt as a continuing goal the exploration of the solar system . . . . To focus our developments and integrate our programs, we recommend that the United States prepare for manned planetary expeditions in the 1980s.[68]

Not surprisingly, the NASA report's program closely resembled the one Paine and von Braun described in their 4 August STG presentation. Continued piloted lunar exploration after Apollo would, the NASA report proclaimed, "expand man's domain to include the [M]oon" by establishing a lunar base. This would lay groundwork for a piloted Mars expedition in the 1980s. As Mayo had proposed, the NASA report described different program rates, each with a different date for reaching Mars, the ultimate goal of all the programs. The "maximum rate," in which money was no object and only the pace of technology could slow NASA's rush to Mars, scheduled the first Mars expedition for 1981. Program I launched the first expedition in 1983, while Program II, the pacing option favored by Agnew, put it in 1986. Program III was identical to Program II, except that no date was specified for the first Mars expedition.

The STG report proper, The Post-Apollo Space Program: Directions for the Future, was also published on 15 September 1969. It had a split personality.[69] The main body closely followed NASA's America's Next Decades in Space report—not surprisingly, since Paine was again the principal author. The introductory "Conclusions and Recommendations" section, however, differed markedly in tone and emphasis from the NASA-authored section. This was because it was added in early September at the insistence of senior White House staffers who did not want to provide President Nixon with only ambitious objectives from which to choose.[70]

The "Conclusions and Recommendations" section acknowledged that NASA had "the demonstrated organizational competence and technology base . . . to carry out a successful program to land man on Mars within 15 years"; however, it failed to advocate an aggressive Mars program, recommending instead sending humans to Mars "before the end of this century." At the same time, it cautioned that "in a balanced program containing other goals and activities, this focus should not assume over-riding priority and cause sacrifice of other important activity in times of severe budget constraints."[71]

New space capabilities would be developed in a three-phase program, to which the introductory section attached no firm schedule. Phase 1 would see "exploitation of existing capability and development of new capability, maintaining program balance within a vailable resources." This would include continued "Apollo-type" lunar missions. New development would be based on the principles of "commonality, reusability, and economy." Phase 2 was an "operational phase" using new systems in cislunar space with emphasis on "exploitation of science and applications" aboard space stations. In Phase 3, "manned exploration missions out of [E]arth-[M]oon space" would occur, "building upon the experience of the ea rlier two phases ."[72] The "Conclusions and Recommendations" section cautioned,

> Schedule and budgetary implications within these three phases are subject to Presidential choice and decision . . . with detailed program elements to be determined in a normal annual budget and program review process. [73]

## Nixon's Response

Shortly after the Apollo 11 lunar landing , von Braun told space policy analyst John Logsdon that

> the legacy of Apollo has spoiled the people at NASA . . . . I believe that there may be too many people in NASA who at the moment are waiting for a miracle, just waiting for another man on a white horse to come and offer us another planet, like President Kennedy.[74]

Von Braun might have placed his boss in that category. Paine placed great stock in the effect the NASA section of the STG report would ha ve on President Nixon. Another document—a lengthy memorandum by Ma yo dated 25 September 1969—apparently had greater effect, however. Mayo told the President that NASA had requested $4.5 billion for FY 1971 despite a $3.5-billion cap imposed by his office. He then recommended that Nixon "hold an announcement of your space deci-

sion until after you have reviewed the [STG] report recommendations specifically in the context of the total 1971 budget problem . . . ." Mayo added that he believed the NASA sections of the STG report "significantly underestimated" the costs of future programs.[75]

In late September, Aviation Week & Space Technology reported that NASA was hopeful that it might receive a supplemental appropriation in FY 1970 to begin work toward Mars.[76] In October this optimism led Mueller to establish the Planetary Missions Requirements Group (PMRG), which included representatives from NASA Headquarters and several NASA field centers . The PMRG, the successor to the Planetary J AG, first met formally in December 1969. Its purpose w as to blueprint Mars mission concepts in the context of the STG integrated plan.[77]

By the time the PMRG met for the first time, however, NASA had received bad news . On 13 November 1969, Mayo's Office of Management and Budget (OMB) (formerly the Budget Bureau) had informed P aine that NASA's FY 1971 request would be $1 billion shy of his request—just $3.5 billion. Paine called the figure "unacceptable" and told Ma yo that "the proposed rationale" for this budget figure "ignores and runs counter to the conclusions reached by the Space Task Group . . . the OMB staff proposals would force the President to reject the Space Program as an important continuing element of his Administration's total program."[78]

Paine was compelled to acquiesce , however. On 13 January 1970, he briefed newsmen on NASA's budget ahead of Nixon's FY 1971 budget speec h. He termed the $3.5 billion budget "solid," and announced that the Saturn V rocket production line , already dormant, would close down permanently.[79] This was a serious blow to the nuc lear rocket program. It meant that, in addition to having no approved mission, it now had no way to get into space. NASA subsequently began study of using the Earth-orbital Space Shuttle to place NERVA-equipped rocket stages into Earth orbit.

Paine also canceled the planned tenth lunar landing mission, Apollo 20, so that its Saturn V could launc h the Skylab space station, and announced that the Viking Mars probe would slip to a 1975 launc h with a 1976 Mars landing. In an apparent effort to raise alarm and fend off further cuts, Paine released a list of NASA Center closures in order of priority. First to go would be

Ames, in Nixon's California stronghold, and the last three in order would be MSC, Marshall, and KSC.[80]

In late January, just before Nixon unveiled his Federal budget for FY 1971, NASA took another cut. When sent to Capitol Hill on 2 February 1970, NASA's portion of the budget had fallen to $3.38 billion. In announcing NASA's budget, Nixon said that "[o]ur actions make it possible to begin plans for a manned mission to Mars"[81] In fact, the 1970-71 period would see NASA 's last formal piloted Mars plan until the 1980s.

Nixon did not use his 22 January 1970 State of the Union address to plot the way forward in space as some in NASA had hoped that he might. His first priority, he said, was to "bring an end to the war in Vietnam." He also proposed to "begin to make reparations for the damage we have done to our air, to our land, and to our waters."[82] Apollo 8 pictures of blue Earth rising over the barren Moon had become a rallying point for the environmental movement—not, as Paine had hoped, for space exploration. Paine was unimpressed by Nixon's environmentalist slant. He told an industry group that "[w]e applaud the increase in sewage disposal plants. But we certainly hope this doesn't mean the nation has taken its eyes off the stars and put them in the sewers."[83]

Nixon finally issued his policy on the post-Apollo space program on 7 March 1970. Unlike Kennedy's 1961 Moon speech, Nixon's statement was broad and vague, with no specifics about NASA funding. Rather than endorse a specific target date for a piloted Mars mission, he said that "we will eventually send men to explore the planet Mars." The British weekly T he Economist reported that people at NASA "looked like children who got the jigsaw puzzle they were expecting rather than the bicycle they were dreaming of."[84]

## PSAC Recommends Shuttle

At the same time Nixon issued his space policy , his PSAC issued T he Next Decade in Space, a report extolling the possibilities of a Space Shuttle-based space program. The presidential advisory body ac knowledged that "[e]normous technological capabilities ha ve been built up in the Apollo Program," but recommended "a civilian space effort about half the magnitude of the present level."[85] The PSAC emphasized the military and direct economic benefits of piloted space travel, which it said could only be accrued by replacing virtually all expendable rockets with a reusable Space Transportation System (STS). This would inc lude the Space Shuttle and a reusable orbital tug .

The STS would allow "orbital assembly and ultimately radical reduction in unit cost of space transportation," the PSAC stated, quoting a NASA/Defense Department study that placed the cost per flight of the STS at $5 million, or 1 percent of the Saturn V cost.[86] At the time the PSAC released its report, the U.S. could launch four Saturn V rockets per year, each with a payload of about 100 tons. The PSAC reasoned that "[s]ince only ten flights of the STS can in principle fulfill the role of two Saturn V launches/year, this capability might be reached soon after initial operation of the STS."[87]

The PSAC then addressed piloted Mars exploration, writing that "[p]rudence suggests that the possibility of undertaking a manned voyage to Mars be kept in mind but that a national commitment to this project be deferred at this time."[88] The STS, it expected, "could place the equipment needed for the Mars mission in orbit with one or two dozen launc hes and at a cost substantially below that of a single Saturn V." It also recommended that the permanent space station it said should precede a piloted Mars mission be deferred until after the STS could be used to assemble it.[89] Despite the heavy reliance it placed on the STS, the PSAC recommended deferring a decision to build it until FY 1972.

In July 1970, Paine submitted his resignation. On 15 September the first anniversary of the release of the STG and NASA reports , George Low took over as Acting NASA Administrator. In February 1971, Presidential Assistant Peter Flanigan w as ordered to find a NASA Administrator who would "turn down NASA's empire-building fervor and turn his attention to . . . work[ing] with the OMB and White House."[90]

## The Last Mars Study

The PMRG, meanwhile, continued low-level Mars expedition planning. NASA's post-Apollo Mars aspirations died with a whimper—a call to NASA Centers participating in the PMRG for reports summing up their work. PMRG work at MSC resided in the Engineering and Development Directorate's Advanced Studies Office

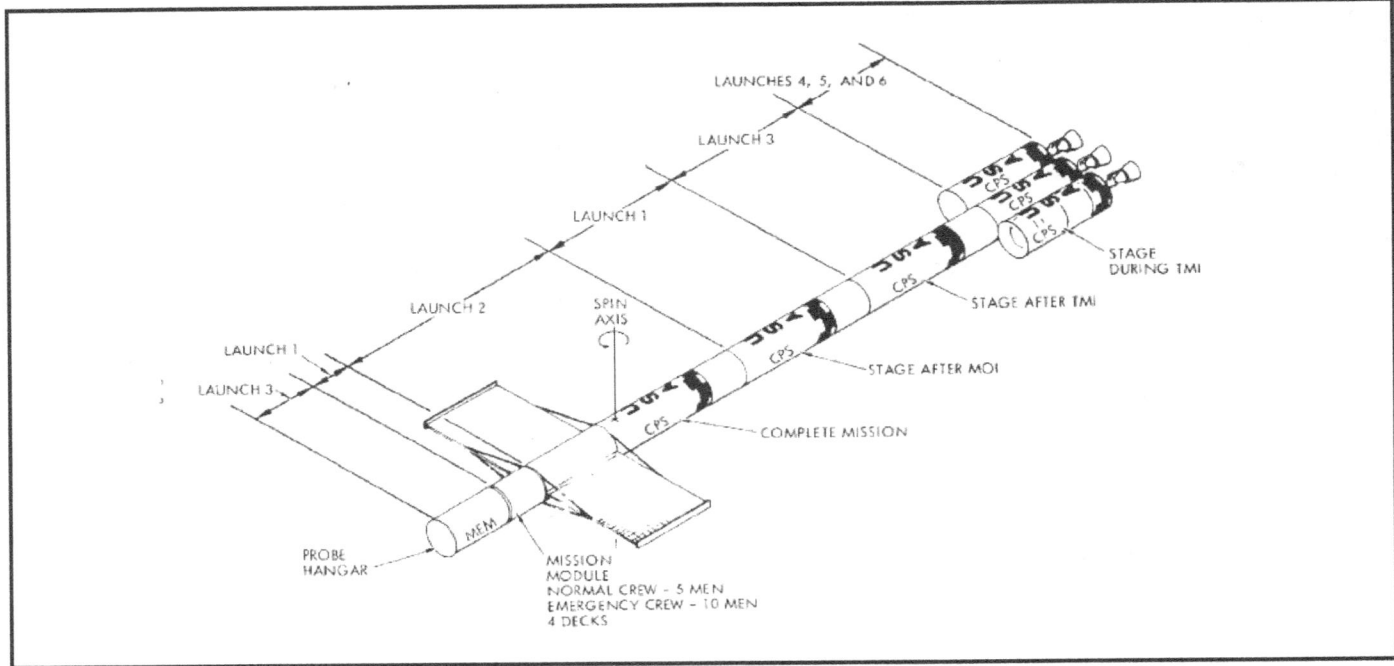

Figure 17—Last gasp (for a while): NASA's 1971 Mars spaceship design, the last until the 1980s, proposed to reduce cost by using projected Space Shuttle tec hnology and rejecting nuclear engines in favor of cheaper chemical propulsion. (Manned Exploration Requirements and Considerations, Advanced Studies Office, Engineering and Development Directorate, NASA Manned Spacecraft Center, Houston, Texas, February 1971, p. 5-2.)

under Morris Jenkins. MSC Associate Director of Engineering Maxime Faget reviewed Jenkins' February 1971 report. In his introduction, Jenkins explained,

> Official statements regarding the manned Mars mission have always been conditioned by an emphasis that there w as no set time frame for it. This together with problems of budget constraints on the more immediate future pro-grams and the overall posture of the space pro-gram, influenced formal support for this study. Justifiably, the formal support was always very small and . . . non-continuous . . . .[91]

The guiding principle of MSC' s PMRG study w as austerity. In general configuration its Mars ship resembled Boeing's 1968 behemoth, but chemical propulsion stood in for nuc lear. According to Jenkins, "everything [was] done to make [this study] a useful point of departure when national priorities and economic considerations encourage the mount-ing of a manned Mars expedition. "[92] MSC targeted its 570-day Mars expedition for the 1987-88 launc h

opportunity, following an 11-year development and test period beginning in the mid-1970s.

MSC assumed a vailability of a fully reusable Space Shuttle based on Max Faget's "flyback" design. The fly-back shuttle would inc lude a winged orbiter launc hed on a winged booster . Both booster and orbiter would carry astronauts. MSC envisioned a booster the size of a 747 and an orbiter on the scale of a DC-9.

The study rejected launching Mars spacecraft compo-nents in the 15-foot-diameter pa yload bay of the orbiter because as many as 30 modules would have to be launched separately and brought together in orbit, necessitating a "complex and lengthy assembly and checkout process."[93] Instead, MSC proposed launc h-ing the Mars ship' s three 24-foot-diameter modules on the bac k of the Shuttle booster with the aid of Chemical Propulsion System (CPS) stages . Three CPS stages would be launc hed into orbit without attached modules.

The Shuttle booster would carry the CPS and attached module (if any) partw ay to orbit, then separate to return to the launch site. The CPS would then ignite to achieve Earth orbit. Each CPS would weigh 30 tons empty and hold up to 270 tons of liquid hydrogen/liquid oxygen propellants. In keeping with the principle of austerity, the CPS stages would use the same rocket engine and propellant tank designs as the Shuttle booster and orbiter, and do double duty as Mars ship propulsion stages. Assembling the expedition's single ship would need 71 Shuttle booster launches. Six would launch the ship (three modules and six CPS stages), and the remainder would carry Shuttle orbiters serving as tankers for loading the CPS stages with propellants.

The assembled Mars ship would include a hangar for automated probes and a MEM based on the 1968 NAR design. For redundancy, its 55-ton, four-deck Mission Module would be split into two independent pressurized volumes, each containing a duplicate spacecraft control station. Deck four would be the ship's solar flare radiation shelter. The 65-foot-long Electrical Power System module would contain pressurized gas storage tanks and twin solar arrays. The crew would rotate the Mars ship end over end about twice per minute to produce artificial gravity in the Mission Module equal to one-sixth Earth's gravity (one lunar gravity).

Earth departure would require a series of maneuvers. Maneuver 1 would expend two CPS stages to place the Mars ship in elliptical "intermediate orbit." Maneuver 2 and Maneuver 3 would use one CPS stage—the first would place the ship in elliptical "waiting orbit," and the second would adjust the plane of the departure path. Space tugs would later recover the three discarded CPS stages for reuse. Maneuver 4 would place the ship on a 6-month trajectory to Mars. The fourth CPS would enter solar orbit after detaching from the Mars ship and would not be recovered.

Slowing the ship so that Mars' gravity could capture it into a 200-mile by 10,000-mile orbit would expend the fifth CPS. The elliptical orbit would require less propellant to enter and depart than a circular one. The five-person crew would spend 15 days in orbit studying Mars and preparing the MEM for landing; then three crewmembers would separate in the MEM, leaving behind two to watch over the mothership.

The MEM crew would explore their landing site using a pair of unpressurized electric rovers resembling the Apollo Lunar Roving Vehicle, which was slated to be driven on the Moon for the first time on Apollo 15 in July 1971. During Mars surface excursions, one crewmember would remain in the MEM while the other two took out one rover each. This "tandem convoy" arrangement would allow the Mars explorers to avoid the "walk back" limit imposed on single-rover traverses in the Apollo program. Walk back distance was limited less by astronaut stamina than by the amount of water and air the space suit backpacks could hold. If one Mars rover failed, the functional rover would return both astronauts to the MEM. Each rover would include a hook for towing the failed rover back to the MEM for repairs.

Rover maximum speed would be 10 miles per hour, and total area available to two rovers would amount to 8,000 square miles, compared to only 80 square miles for a single rover. Once every 15 days, a 36-hour traverse of up to 152 miles would occur, with the astronauts sleeping through the frigid Martian night on the parked rovers in their hard-shelled aluminum space suits. Jenkins did not attempt to estimate the amount of sleep the astronauts might actually be able to achieve during their overnight camping trips.

The astronauts would collect samples of rock and soil with emphasis on finding possible life. According to the MSC report, "[t]he potential for even elementary life to exist on another planet in the solar system may . . . be the keystone to the implementation of a manned planetary exploration program . . . man's unique capabilities in exploration could . . . have a direct qualitative impact on life science yield."[94]

After 45 days of surface exploration, the crew would blast off in the MEM ascent stage and dock with the mothership. Any specimens of Mars life collected would be transferred to a Mars environment simulator. The crew would discard the ascent stage; then the sixth and final CPS would ignite to push the ship back toward Earth. The MEM astronauts would remain quarantined in one pressurized volume until the danger of spreading Martian contagion to the other astronauts was judged to be past.

The MSC PMRG report received only limited distribution within NASA and virtually none outside the Agency. Formal studies within NASA aimed at sending humans to Mars would not occur again until the Manned Mars Missions exercise in 1984 and 1985.

## Chapter 5: Apogee

## NERVA Falls, Shuttle Rises

The OMB's FY 1972 request for NASA w as $3.31 billion. The budget slashed NERV A funding in fa vor of continued Space Shuttle studies . Combined AEC-NASA nuclear rocket funding plummeted to $30 million split evenly between the two agencies . NASA and the AEC had together requested $110 million. The allotted budget threatened to place the NRDS on standby and was considered by many sufficient only to shut down the program.

In February 1971, Clinton Anderson held a hearing on the cut in NASA's NERVA funding. In his introductory remarks, he lauded the nuclear rocket program as "one of the most successful space technology programs ever undertaken" and pointed to the $1.4-billion investment in nuclear propulsion technology since 1955.[96] Senator Alan Bible (Democrat-Nevada) then pointed out that the STG report called for nuclear rockets.[96]

Acting NASA Administrator George Low took h is marching orders from the highest levels of the Nixon White House. The Earth-orbital Shuttle had to come first, he said—without it NERVA had no ride to space . He told the Senators that, "NERVA needs the Shuttle, but the Shuttle does not need NERVA."[97]

Low denied that the funding cut would kill the program, explaining that "useful work on long lead-time items" could be accomplished.[98] There would, however, be no technical progress during FY 1972, and possibly none in FY 1973. "We have not, as yet, been able to look forward beyond that," Low added.[99]

Two months later , in May 1971, 21 members of Congress wrote to President Nixon requesting more funds for NERVA in FY 1972. When the White House failed to respond, Congress of its own accord budgeted

$81 million for nuc lear rockets, of which NASA's portion was $38 million. In October, however, the OMB refused to release more than the $30 million the Administration had requested. In November the OMB stood by its FY 1972 nudear propulsion request despite protests from the Senate floor.[100]

On 5 January 1972, President Nixon met with J ames Fletcher, Tom Paine's successor as NASA Administrator, at the "Western White House" in San Clemente, California. Afterward, Fletcher read out Nixon's statement calling for an FY 1973 new start on the Shuttle. The announcement's venue w as significant—California, a state of many aerospace firms, was vital to Nixon's 1972 reelection bid.[101] Nixon pointed out that "this major new national enterprise will engage the best efforts of thousands of highly skilled workers and hundreds of contractor firms over the next several years." Fletcher added that it was "the only meaningful new manned program that can be accomplished on a modest budget."[102] First flight was scheduled for 1978.

Nixon sent his FY 1973 budget to Capitol Hill on 24 January 1972. As its supporters had feared, the budget contained no funds for NERVA. Anderson, nuclear propulsion's greatest c hampion, was ill and could not defend it. The last NERVA tests occurred in J une and July of 1972. Anderson retired from the Senate at the end of 1972. The FY 1974 budget terminated what remained of the U.S. nuclear rocket program.[103]

With both NERVA and Saturn V gone—the last Saturn V flew in May 1973—NASA's piloted space flight ambitions collapsed back to low-Earth orbit. Yet the Agency did not cease to strive tow ard Mars. As we will see in the next chapter, NASA's robot explorers conducted the first in-depth Mars exploration in the 1970s , holding open the door for renewed piloted Mars planning .

# Chapter 6: Viking and the Resources of Mars

Additional automated missions will most certainly occur, but the ultimate scientific study of Mars will be realized only with the coming of man—man who can conduct seismic and electromagnetic sounding surveys; who can launch balloons, drive rovers, establish geologic field relations, select rock samples and dissect them under the microscope; who can track clouds and witness other meteorological transients; who can drill for permafrost, examine core tubes, and insert heat-flow probes; and who, with his inimitable capacity for application of scientific insight and methodology, can pursue the quest for indigenous life forms and perhaps discover the fossilized remains of an earlier biosphere. (Benton Clark, 1978)[1]

## The New Mars

In the 1960s, most automated missions beyond low-Earth orbit—the Rangers, Surveyors, and Lunar Orbiters—supported the piloted Apollo program. In the 1970s, as NASA's piloted program contracted to low-Earth orbit, its automated program expanded beyond the Moon. Sophisticated robots flew by Mercury, Jupiter, and Saturn, and orbited and landed on Venus and Mars.

Though they were not tailored to serve as precursors to human expeditions in the manner of the Rangers, Surveyors, and Lunar Orbiters, the automated missions to Mars in the 1970s shaped the second period of piloted Mars mission planning, which began in about 1981. The first of these missions, Mariner 9, took advantage of the favorable Earth-Mars transfer opportunity associated with the August 1971 opposition to carry enough propellant to enter Mars orbit. It was launched from Cape Kennedy on 30 May 1971.

In September, as Mariner 9 made its way toward Mars, Earth-based astronomers observing the planet through telescopes saw a bright cloud denoting the onset of a dust storm. By mid-October it had become the largest on record. Wind-blown dust obscured the entire surface, raising fears that Mariner 9 might not be able to map the planet from orbit as planned.[2]

On 14 November 1971, after a 167-day Earth-Mars transfer, Mariner 9 fired its engine for just over 15 minutes to slow down and become Mars' first artificial satellite. Dust still veiled the planet, so mission controllers pointed the spacecraft's cameras at the small Martian moons Phobos and Deimos. In Earth-based telescopes they were mere dots nearly lost in Mars' red glare. In Mariner 9 images, Phobos was marked by parallel cracks extending from a large crater. Apparently the impact that gouged the crater had nearly smashed the little moon. Deimos, Mars' more distant satellite, had a less dramatic, dustier landscape.

The giant dust storm subsided during December, theatrically unveiling a surprising world. Mars was neither the dying red Earth espoused by Percival Lowell nor the dead red moon glimpsed by the flyby Mariners.[3] From its long-term orbital vantage point, Mariner 9 found Mars to be two-faced, with smooth northern lowlands and cratered southern highlands. The missions to the Moon confirmed that a relationship exists between crater density and age—the more densely cratered a region, the older it is. Hence, Mars has an ancient hemisphere and a relatively young hemisphere.

Mars is a small world—half Earth's diameter—with large features. The Valles Marineris canyons, for example, span more than 4,000 kilometers along Mars' equator. Nix Olympica, imaged by Mariner 6 and Mariner 7 from afar and widely interpreted as a bright crater, turned out to be a shield volcano 25 kilometers tall and 600 kilometers wide at its base. Renamed Olympus Mons ("Mount Olympus"), it stands at one edge of the Tharsis Plateau, a continent-sized tectonic bulge dominating half the planet. Three other shield volcanoes on the scale of Olympus Mons form a line across Tharsis' center.

Most exciting for those interested in Martian life were signs of water. Mariner 9 charted channels tens of kilometers wide. Some contain streamlined "islands" apparently carved by enormous rushing floods. Many of the giant channels originate in the southern highlands and open out onto the smooth northern plains. The northern plains preserve rampart craters—also called "splosh" craters—which scientists believe were formed by asteroid impacts in permafrost. The heat of impact apparently melted subsurface ice, which flowed outward from the impact as a slurry of red mud, then refroze.[4]

# Chapter 6: Viking and the Resources of Mars

Mariner 9 depleted its nitrogen attitude-control propellant on 27 October 1972, after returning more than 7,200 images to Earth. Controllers quickly lost radio contact as it tumbled out of control. A week later, on 6 November 1972, mission planners using Mariner 9 images announced five candidate Viking landing sites.[5]

Viking 1 left Earth on 20 August 1975 and arrived in Mars orbit on 19 June 1976. Its twin, Viking 2, left Earth on 9 September 1975 and arrived at Mars on 7 August 1976. The spacecraft consisted of a nuclear-powered lander and a solar-powered orbiter. The Viking 1 lander separated from its orbiter and touched down successfully in eastern Chryse Planitia on 20 July 1976. Viking 2 alighted near the crater Mie in Utopia Planitia on 3 September 1976.

The first color images from the Viking 1 lander showed cinnamon-red dirt, gray rocks, and a blue sky. The sky color turned out to be a processing error based on preconceived notions of what a sky should look like. When the images were corrected, Mars' sky turned dusky pink with wind-borne dust.[6]

The Vikings confirmed the old notion that Mars is the solar system planet most like Earth, but only because the other planets are even more alien and hostile. A human dropped unprotected on Mars' red sands would gasp painfully in the thin carbon dioxide atmosphere, lose consciousness in seconds, and perish within two minutes. Unattenuated solar ultraviolet radiation would blacken the corpse, for Mars has no ozone layer. The body would freeze rapidly, then mummify as the thin, parched atmosphere leeched away its moisture.

By the time the Vikings landed, almost no one believed any longer that multicellular living things could exist on Mars. They held out hope, however, for hardy single-celled bacteria. On 28 July 1976, the Viking 1 lander scooped dirt from the top few centimeters of Mars' surface and distributed it among three exobiology detectors and two spectrometers. The instruments returned identical equivocal readings—strong positive responses that tailed off, weak positive responses that could not be duplicated in the same sample, and, most puzzling, an absence of any organic compounds the instruments were designed to detect.

Viking 1 and Viking 2 each scooped additional samples—even pushing aside a rock to sample underneath—and repeated the tests several times with similar equivocal results. Most scientists interpreted the Viking results as indicative of reactive soil chemistry produced by ultraviolet radiation interactions with Martian dirt, not of life. The reactive chemistry probably destroys any organic molecules.[7]

Improved cameras on the Viking orbiters, meanwhile, added detail to Mariner 9's Mars map. They imaged polygonal patterns on the smooth northern plains resembling those formed by permafrost in Earth's Arctic regions. Some craters—Gusev, for example—looked to be filled in by sediments and had walls breached by sinuous channels. Perhaps they once held ice-clad lakes.

The Viking images also revealed hundreds of river-size branching channels—called "valley networks"—in addition to the large outflow channels seen in Mariner 9 images. Though some were probably shaped by slowly melting subsurface ice, others appeared too finely branched to be the result of anything other than surface runoff from rain or melting snow. Ironically, most of the finely branched channels occurred in the southern hemisphere, the area that reminded people in the 1960s of Earth's dead Moon. The flyby Mariners might have glimpsed channels among the Moonlike craters had their cameras had better resolution.

Low pressure and temperature make free-standing water impossible on Mars today. The channels in the oldest part of Mars, the cratered southern highlands, seem to point to a time long ago when Mars had a dense, warm atmosphere. Perhaps Mars was clement enough for a sufficiently long period of time for life to form and leave fossils.[8]

The Viking landers and orbiters were gratifyingly long-lived. The Viking 1 orbiter functioned until 7 August 1980. Together with the Viking 2 orbiter, it returned more than 51,500 images, mapping 97 percent of the surface at 300-meter resolution. Though required to operate for only 90 days, the Viking 1 lander, the last survivor of the four vehicles, returned data for more than six years. The durable robot explorer finally broke contact with Earth on 13 November 1982.[9]

Viking was a tremendous success, but it had been widely billed as a mission to seek Martian life. The inconclusive Viking exobiology results and negative interpretation placed on them helped dampen public enthusiasm for Mars exploration for a decade. Yet Viking showed Mars to be eminently worth exploring. Moreover, Viking revealed abundant resources that might be used to explore it.

## Living off the Land

During the period that Mariner 9 and the Vikings revealed Mars to be a rich destination for explorers, almost no Mars expedition planning occurred inside or outside NASA. The Agency was preoccupied with developing the Space Shuttle, and Mars planners independent of NASA—who would make many contributions during the 1980s—were not yet active in significant numbers.

Papers on In-Situ Resource Utilization (ISRU) were among the first signs of re-awakening interest in piloted Mars mission planning. ISRU is an old concept, dating on Earth to prehistory. ISRU can be defined as using the resources of a place to assist in its exploration—the phrase "living off the land" is essentially synonymous. In the context of space exploration, ISRU enables spacecraft weight minimization. If a spacecraft can, for example, collect propellants at its destination, those propellants need not be transported at great expense from Earth's surface. In the 1960s, ISRU was studied largely in hopes of providing life-support consumables. By the 1980s, the propellant production potential of ISRU predominated.

NASA first formally considered ISRU in 1962, when it set up the Working Group on Extraterrestrial Resources (WGER). The WGER, which met throughout the 1960s, focused on lunar resources, not Martian. This was because more data were available on lunar resource potential, and because lunar resource use was, in the Apollo era, potentially more relevant to NASA's activities.[10]

The UMPIRE study (1963-1964) recommended applying ISRU to establish and maintain a Mars base during long conjunction-class surface stays. Doing this would, of course, demand more data on what resources were available on Mars. NASA Marshall's UMPIRE summary report stated that "[t]his information, whether it is obtained by unmanned probes or by manned [flyby or orbiter] reconnaissance missions, would

make such a base possible," making the " 'cost effectiveness' of Mars exploration . . . much more reasonable than [for] the short excursions."[11]

Fifteen years after UMPIRE, the Vikings at last produced the in-situ data set required for serious consideration of Mars ISRU. The first effort to assess the potential of Martian propellant production based on Viking data spun off a 1977-78 NASA JPL study of an automated Mars sample-return mission proposed as a follow-on to the Viking program. Louis Friedman headed the study, which was initially inspired by President Gerald Ford's apparently casual mention of a possible "Viking 3" mission soon after the successful Viking 1 landing.[12] Robert Ash, an Old Dominion University professor working at JPL, and JPL staffers William Dowler and Giulio Varsi published their results in the July-August 1978 issue of the refereed journal Acta Astronautica.[13]

They examined three propellant combinations. Liquid carbon monoxide and liquid oxygen, they found, were easy to produce from Martian atmospheric carbon dioxide, but they rejected this combination because it produced only 30 percent as much thrust as liquid hydrogen/liquid oxygen. Electrolysis (splitting) of Martian water could produce hydrogen/oxygen, but they rejected this combination because heavy, energy-hungry cooling systems were necessary to keep the hydrogen liquid, thus negating the weight-reduction advantage of in-situ propellant manufacture.

Liquid methane/liquid oxygen constituted a good compromise, they found, because it yields 80 percent of hydrogen/oxygen's thrust, yet methane remains liquid at higher temperatures, and thus is easier to store. The Martian propellant factory would manufacture methane using a chemical reaction discovered in 1897 by French chemist Paul Sabatier. In the Sabatier reaction, carbon dioxide is combined with hydrogen in the presence of a nickel or ruthenium catalyst to produce water and methane. The manufacture of methane and oxygen on Mars would begin with electrolysis of Martian water. The resultant oxygen would be stored and the hydrogen reacted with carbon dioxide from Mars' atmosphere using the Sabatier process. The methane would be stored and the water electrolyzed to continue the propellant production process.

Ash, Dowler, and Varsi estimated that launching a one-kilogram sample of Martian soil direct to Earth would need 3.8 metric tons of methane/oxygen, while launching a piloted ascent vehicle into Mars orbit would need 13.9 metric tons. These are large quantities of propellant, so conjunction-class trajectories with Mars surface stay-times of at least 400 days would be necessary to provide enough time for propellant manufacture.

Benton Clark, with Martin Marietta (Viking's prime contractor) in Denver, published the first papers exploring the life-support implications of the Viking results. His 1978 paper entitled "The Viking Results—The Case for Man on Mars" pointed out that every kilogram of food, water, or oxygen that had to be shipped from Earth meant that a kilogram of science equipment, shelter structure, or ascent rocket propellant could not be sent.[14] Clark estimated that supplies for a 10-person, 1,000-day conjunction-class Mars expedition would weigh 58 metric tons, or about "one hundred times the mass of the crew-members themselves." The expedition could, however, reduce supply weight, thereby either reducing spacecraft weight or increasing weight available for other items, by extracting water from Martian dirt and splitting oxygen from Martian atmospheric carbon dioxide during its 400-day Mars surface stay.

Clark wrote that Mars offered many other ISRU possibilities, but that they probably could not be exploited until a long-term Mars base was established. This was because they required structures, processing equipment, or quantities of power unlikely to be available to early expeditions. Crop growth using the "extremely salty" Martian soil, for example, would probably have to await availability of equipment for "pre-processing . . . to eliminate toxic components."[15]

The Vikings' robotic scoops barely scratched the Martian surface, yet they found useful materials such as silicon, calcium, chlorine, iron, and titanium. Clark pointed out that these could supply a Mars base with cement, glass, metals, halides, and sulfuric acid. Carbon from atmospheric carbon dioxide could serve clever Martians as a foundation for building organic compounds, the basis of plastics, paper, and elastomers. Hydrogen peroxide made from water could serve as powerful fuel for rockets, rovers, and powered equipment such as drills.

During the 1980s, the Mars ISRU concept generated papers by many authors, as well as initial experimentation.[16] Robert Ash, for example, developed experimental Mars ISRU hardware at Old Dominion University with modest funding support from NASA Langley[17] and from a non-government space advocacy group, The Planetary Society.[18] That a private organization would fund such work was significant.

Before ISRU could make a major impact, piloted Mars mission planning had to awaken more fully from its decade-long post-Apollo slumber. Post-Apollo Mars planning occurred initially outside official NASA auspices. This constituted a sea-change in Mars planning—up to the 1970s, virtually all Mars planning was government-originated. In the 1980s, as will be seen in the coming chapters, individuals and organizations outside the government took on a central, shaping role.

# Chapter 7: The Case for Mars

We didn't know all of the people who finally did speak . . . until they called us! Somehow they heard about the conference, through the flyers we sent around and from word of mouth, and they volunteered. It really w as a Mars Underground! (Christopher McKay, 1981)[1]

## Columbia

Columbia, the first Space Shuttle , lifted off from P ad 39A at Kennedy Space Center on 12 April 1981, with Commander John Young and Pilot Robert Crippen on board for a two-day test flight. Nearly 12 years before, the Apollo 11 CSM Columbia had left the same pad atop a Saturn V at the start of the first Moon landing mission. For Shuttle flights, the twin Complex 39 pads were trimmed back and heavily modified. Designated STS-1, it was the first U.S. piloted space flight since the joint United States-Soviet Apollo-Soyuz mission in July 1975.

At launch, the 2,050-metric-ton Shuttle "stack" consisted of the delta-winged orbiter Columbia and twin 45.4-meter-long Solid Roc ket Boosters (SRBs) attached to a 47.4-meter -long expendable External Tank (ET). Columbia measured 37.2 meters long with a wingspan of 23.8 meters . Seconds before planned liftoff, the three Space Shuttle Main Engines (SSMEs) in the orbiter's tail ignited in rapid sequence, drawing liquid hydrogen and liquid oxygen propellants from the ET. Then, at T-0, the two SRBs lit up . Unlike the Saturn V, which climbed slowly during first-stage operation, Columbia leapt from the launc h pad. Also unlike the Saturn V engines, the SRBs could not be turned off once they ignited, making abort impossible until they exhausted their propellants and separated. This was not considered a major risk—SRBs , used since the 1950s, were considered a mature technology.

Two minutes into STS-1, the SRBs separated and fell into the Atlantic for recovery and reuse. Columbia's SSMEs, the world's first reusable large roc ket engines, continued pushing the orbiter and ET tow ard space. Eight and one-half minutes after launc h, the SSMEs shut down and the ET separated. Young and Crippen fired Columbia's twin Orbiter Maneuvering System (OMS) engines to complete orbital insertion while the ET tumbled and reentered, then opened the long doors covering Columbia's 18.3-meter by 4.6-meter payload bay.

The payload bay was the orbiter's raison d'être. Maximum payload to low-Earth orbit w as about 30 metric tons, though center of gra vity and landing weight constraints restricted this to some degree. The payload bay could carry satellites for release into orbit or a European-built pressurized laboratory module called Spacelab. The Space Shuttle orbiter was also the only space vehicle that could rendezvous with a satellite and capture it for repair or return to Earth—it could return about 15 metric tons to Earth in its pa y-load bay. NASA hoped to use the Space Shuttle to launch components for an Earth-orbiting space station and other vehic les, such as aerobraking Orbital Transfer Vehicles (OTVs) based at the station.

On 14 April Young and Crippen fired Columbia's OMS engines for about two minutes to begin reentry . The STS-1 reentry had almost nothing in common with previous piloted reentries. Columbia's heat shield did not ablate—that is, burn away—to protect it from the friction heat of reentry. Instead, in the interest of reusability, Columbia relied on more than 24,000 individually milled spun-glass tiles to shield its aluminum skin.

After a pair of c lose-timed sonic booms—they would become a Space Shuttle trademark—Columbia glided to a touchdown on the wide dry lake bed at Edw ards Air Force Base, California. Future landings would occur on a runway seven kilometers from the Complex 39 Shuttle pads at KSC.[2]

NASA heralded the flight as the start of a new era of routine, inexpensive space access that might spa wn industry off Earth. An ebullient Young told reporters, "We're not really too far—the human race isn't—from going to the stars."[3]

## The Case for Mars

The Mars buffs who were gathered i n Boulder, Colorado, for the first Case for Mars conference , just two weeks after the first Shuttle flight, would have settled for NASA's setting its sights on Mars—never mind the stars. Fueled by the Viking discoveries, would-be Mars explorers dared look beyond the Space Shuttle. They hoped that Mars ship propellants and components might soon be manifested as Shuttle pa yloads. They also saw in the Shuttle and in the Space Station Program (expected soon to follow) sources of hardw are for Mars ship parts, much as planners in the 1960s envisioned using Apollo hardware for piloted Mars flybys.

The 1981 Case for Mars conference provided the first public forum for Mars planning since the 1960s . The conference crystallized around an informal seminar based on NASA's 1976 study The Habitability of Mars, organized by Christopher McKay, a University of Colorado at Boulder astro-geophysics Ph.D. candidate. The seminar brought together Mars enthusiasts from Boulder and around the country . The "Mars Underground," as they light-heartedly called themselves, decided in the spring of 1980 that the time w as ripe for a conference on Mars exploration. [4]

The Case for Mars conference drew its name from the title of Benton Clark's 1978 Mars ISRU paper (see chapter 6). Clark took part in the conference , along with about 300 other engineers , scientists, and enthusiasts.[5] It was the largest gathering of would-be Mars explorers since the 1963 Symposium on the Manned Exploration of Mars.

The conference was in part a brain-storming session— an opportunity to take stock of ideas on how to explore Mars. Among the concepts presented was S. Fred Singer's "PH-D Proposal," which drew upon Shuttle-related technology expected to exist in 1990. [6] Singer's scenario had staying power—he was still writing about it in the spring of 2000.[7]

Singer's $10-billion expedition would use Deimos, Mars' outer moon, as a base of operations for exploring the Martian system. It was similar to the 1960s piloted flyby and orbiter missions in how it minimized spacecraft weight. None of the six to eight astronauts would land on Mars , though a sample-return lander would bring up a "grab sample" from the planet and two astronauts would visit Phobos, Mars' inner moon. The astronauts would remote-control between 10 and 20 Mars surface rovers during their two-to-six-month stay in the Martian system. At Deimos' orbital distance, round-trip radio travel time would be only one-fifth of a second.

Two astronauts would be "medical scientists" who would study human reactions to weightlessness, radiation, and isolation throughout the expedition. They would minimize risk to the crew from these long-duration space flight hazards by continually monitoring their health; data they gathered would also minimizerisk for future Mars landing expeditions .

Singer's expedition would r ely on s olar-electric thrusters, using electricity from a large solar arra y to ionize and electrostatically expel argon gas . As described in c hapter 2, electric propulsion thrusters produce constant low-thrust acceleration while using very little propellant. Singer assumed that the solar array would be available in high-Earth orbit in 1990 as part of a pre-existing Shuttle-launc hed cislunar infrastructure. The cost of the solar arra y was thus not included in Singer's expedition cost estimate.

At the start of the PH-D Proposal expedition, the unpiloted solar-electric propulsion system would spiral out from Earth, slowly gaining speed. Several weeks later, as it was about to escape Earth orbit, the piloted Phobos-Deimos craft would catc h up, using chemical rockets, and dock. This technique minimized risk to crew by reducing the amount of time they had to spend in weightlessness and by speeding them through the Van Allen Radiation Belts. The solar-electric propulsion system would accelerate the spacecraft until expedition mid-point; then its thrusters would be turned to point in the direction of travel. The spacecraft would then decelerate until it was captured into orbit by Mars' gra vity.

The 1990-91 target launch date would allow Singer' s expedition to take advantage of a Venus flyby opportunity to gain speed and change course without using propellant. Total expedition duration would be "something less than two years ." Electric propulsion plus Venus flyby plus postponing the piloted Mars landing until a later expedition would reduce spacecraft weight at Earth-orbit departure to about 300 tons .

## The Planetary Society

In 1983, The Planetary Society, a non-profit space advocacy organization with about 100,000 members, commissioned the most detailed piloted Mars mission study since 1971. The organization did this because, as Society president Carl Sagan and executive director Louis Friedman wrote in their foreword to the study report, "since Apollo, there have been, in the United States at least, almost no serious studies of manned (or womanned) voyages to other worlds , despite the fact that enormous technological advances have been made since those early lunar landings."[8] Writing in the Society's member magazine, The Planetary Report, Friedman explained that "we funded [the study] because it is important to have solid technical evidence to back us in our advocacy of new goals . . . ."[9] The nine-month study, "a labor of love" performed at a "bargain basement price" by Science Applications International Corporation (SAIC), was completed in September 1984.[10]

SAIC's Mars mission design resembled MSC's 1963 Flyby-Rendezvous mode. Eighteen Space Shuttle launches would deliver more than 160 metric tons of spacecraft components to Earth orbit. The four-person crew would travel to Mars in a 121-metric-ton Outbound Vehicle consisting of four "sub-vehicles." These were the 38-metric-ton Interplanetary Vehicle, the 19-metric-ton Mars Orbiter, the 54-metric-ton Mars Lander, and the 10-metric-ton Mars Departure Vehicle.

The Interplanetary Vehicle, which would provide one-quarter of Earth's gravity by spinning three times each minute, would include pressurized crew modules based on Spacelab modules. The Mars Orbiter, the Mars Departure Vehicle, and the conical, two-stage Mars Lander were together designated the Mars Exploration Vehicle (MEV). The MEV would include a 54-meter-diameter aerobrake. The crew would return from Mars in the 43-metric-ton Earth Return Vehicle (ERV), which resembled the Interplanetary Vehicle except in that it would include a conical 4.4-metric-ton Earth-return capsule nested in a 13-meter-diameter aerobrake. Of these vehicles, only the MEV would have to slow down and enter Mars orbit. This, plus extensive use of aerobraking, would reduce the amount of propellant required to carry out SAIC's Mars expedition, which in turn would reduce spacecraft weight.

The unpiloted ERV would depart Earth orbit on 5 June 2003, using three large OTVs stacked together, each carrying over 27 metric tons of propellants. The SAIC team assumed that reusable space-based OTVs would be available in Earth orbit as part of NASA's Space Station Program. The expense of developing, launching, and operating the OTVs was thus not counted in the cost of the expedition. OTV 1 would fire its engines at perigee, increasing its apogee distance, then separate. OTV 2 would repeat this procedure. OTV 3's perigee burn would place the ERV on course for Mars. This series of maneuvers would require about six hours.

The crew would depart Earth in the Outbound Vehicle ten days later, on 15 June 2003. Because it was nearly three times heavier than the ERV, the Outbound Vehicle would need perigee burns by seven OTVs over about two days to achieve a Mars-bound trajectory.

On 24 December 2003, the crew would near Mars in the Outbound Vehicle, board and undock the MEV, and aerobrake into Mars orbit. The abandoned Interplanetary Vehicle would fly past Mars into solar orbit. Three of the four crew would enter the Mars Lander and descend to the surface, which they would explore using a pressurized rover. On 23 January 2004, after a month on Mars, the surface crew would lift off in the Mars Lander ascent stage with 400 kilograms of rock samples to rejoin their colleague aboard the Mars Orbiter.

The ERV, meanwhile, would approach Mars on a flyby trajectory. The crew would board the Mars Departure Vehicle, abandon the Mars Lander ascent stage and Mars Orbiter, and leave Mars orbit in pursuit of the ERV. Rendezvous and docking would occur while the ERV was outbound from Mars. Friedman noted that "[b]ecause the orbit doesn't close around Mars, the crew has only a chance at one precise time, to rendezvous with the return vehicle. Although this is risky, SAIC analysis found it acceptable compared to other mission risks. (However, some of us at The Planetary Society wonder if the crew will feel the same way!)"[11]

Eighteen months later, on 5 June 2006, the crew would board the Earth-return capsule and separate from the ERV. They would aerobrake in Earth's atmosphere while the abandoned ERV flew past Earth into solar orbit.

The SAIC team estimated the cost of their Mars expedition at $38.5 billion in 1984 dollars, of which $14.3 billion would be spent on Mars spacecraft hardware, $2 billion would pay for Shuttle launches, and $18.5 billion would be spent on operations. Friedman pointed out that, over a decade, this cost averaged about $4 billion annually, or about 60 percent of NASA's approximately $7-billion FY 1984 budget.[12]

# Chapter 7: The Case for Mars

## Space Station

Traditionally, space stations have been envisioned as having multiple functions, not least of which was as an assembly and servicing base—a spaceport—for spacecraft, including those bound for Mars. In 1978, as Space Shuttle development moved into its final stages, NASA's Johnson Space Center (JSC) (as MSC was renamed in 1973) began planning a Shuttle-launched modular space station called the Space Operations Center (SOC). A space shipyard, the SOC was the most important station concept in the years 1978 through 1982—the period immediately before gaining approval for a station became a realistic goal for NASA. [13]

An internal NASA presentation in May 1981—one month after STS-1—described the SOC as the central element of a "space operations system" that would include the Space Shuttle, OTVs for moving objects assembled at the SOC to orbits beyond the Shuttle's altitude limit, and Manned OTVs for transporting astronauts on satellite service calls. [14] A JSC press release in early 1982 referred to the SOC as a "space base and marshaling yard for large and complex payloads" providing "garage space for reusable cryogenic stages." [15]

After his first year in office, during which NASA took deep cuts, President Ronald Reagan came to see the political benefits of being identified with a successful space program. On 4 July 1982, Columbia returned from space at the end of mission STS-4. Reagan was on hand amid fluttering American flags to declare the Shuttle operational. He spoke of establishing a more permanent presence in space, but withheld a clear mandate to build a space station until his 25 January 1984 State of the Union address. When he did, he emphasized its laboratory function:

> We can follow our dreams to distant stars, living and working in space for peaceful economic and scientific gain. Tonight I am directing NASA to develop a permanently manned Space Station and to do it within a decade. The Space Station would permit quantum leaps in our research in science, communications, in metals and in life-saving medicines which can only be manufactured in space . . . . [16]

The lab function was emphasized partly to keep the Station's estimated cost as close to $8 billion as possible. [17] As we have seen, Mars planners early in the 1980s assumed that OTVs and other Earth-orbit infrastructure applicable to Mars exploration would soon become available. With the spaceport role de-emphasized and the lab role moved to the fore, the justification for OTVs was largely removed, and the ability to assemble other Earth-orbit infrastructure, such as Singer's solar array, was made forfeit. Assembling the Space Station itself would provide some experience with application to Mars ship assembly. However, it would provide little experience with handling tankage and propellants in space, both crucial to building a Mars ship.

## Soviets to Mars

In the early 1980s, such NASA advanced planning as existed focused more on the Moon than on Mars. The revival in NASA Mars interest owes much to geologist and Apollo 17 Moonwalker Harrison Schmitt, and to the Agency's lunar base studies, which had never receded to the same degree as its Mars studies. Schmitt was concerned about an on-going Soviet space buildup, which saw long stays by cosmonauts aboard Salyut space stations and development of a Soviet shuttle and heavy-lift rocket, as well as plans for ambitious robotic Mars missions. [18] Schmitt also concentrated on Mars because he had asked children, the future space explorers, about returning to the Moon and found that they were not interested because people had already been there. [19]

Schmitt had attempted to promote Mars exploration in the late 1970s while serving as Republican U.S. Senator from New Mexico. Following Viking's success, he had put forward a bill calling for the U.S. to develop the capability to establish a settlement on Mars by 2010. His "Chronicles Plan" excited momentary interest in President Jimmy Carter's White House, inducing NASA Administrator Robert Frosch to activate a small NASA study team. The team's July 1978 workshop at Wallops Island, Virginia, produced nothing new. In fact, the consensus was that "past work [from the 1960s] should not be updated unless serious consideration is being given to conducting a manned Mars mission prior to the year 2000." [20] In short, NASA was too busy working on the Space Shuttle in 1978 to think about Mars.

Schmitt renewed his Mars efforts in 1983 by contacting Paul Keaton of LANL during a meeting held in the run-up to the 1984 Lunar Bases and Space Activities of the 21st Century conference, held at the National Academy of Sciences in Washington, DC.[21] As seen in Chapter 5, LANL was involved in space flight before NASA w as created through its work on nuclear rockets. At the lunar base conference, Schmitt presented a paper on his "Mars 2000 Millennium Project, " which he hoped would "mobilize the energies and imaginations of young people who are already looking beyond Earth orbit and the [M]oon. "[22] He also made contact with NASA engineers and scientists interested in exploring Mars as well as the Moon.[23]

Schmitt then pressed for a study to give the President the option to send humans to Mars should he desire to respond to the Soviet buildup . LANL partnered with NASA to conduct the Manned Mars Mission (MMM) study during 1984 and 1985. The effort culminated in the joint NASA-LANL MMM workshop at NASA Marshall (10-14 June 1985).[24] The workshop published three volumes of proceedings in 1986.[25]

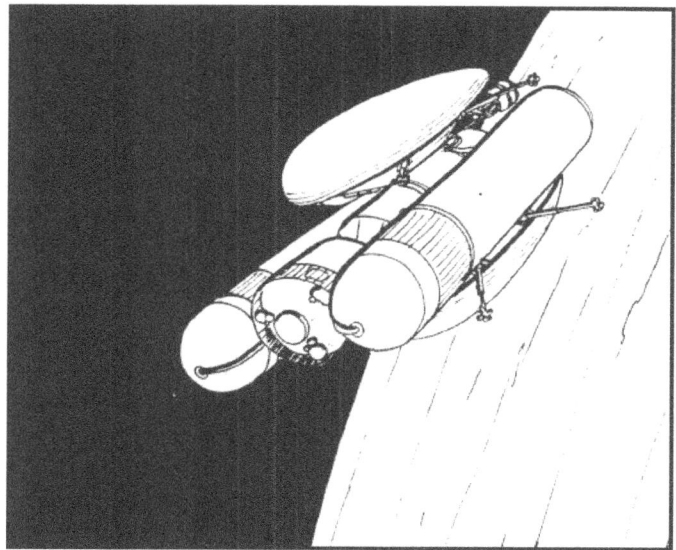

Figure 18—In 1985, NASA's Johnson Space Center responded to suspected Sovie t Mars plans by proposing a U.S. Mars flyby using Space Station and lunar base hardware then planned for the 1990s . Here the flyby spacecraft orbits Earth bef ore setting out fo r Mars. ("Concept for a Manned Mars Flyby," Barney Roberts, Manned Mars Missions: Working Group Papers, NASA M002, NASA/Los Alamos National Laboratories , Huntsville, Alabama/Los Alamos, New Mexico, June 1986, Vol. 1, p. 210.)

Especially noteworthy, given Schmitt's primary rationale for the MMM workshop, was a JSC plan for a piloted Mars flyby based on technology expected to exist in the 1990s as part of the Space Station Program. This aimed at countering a possible Soviet piloted Mars flyby mission.

In April 1985, at Schmitt's request, the CIA prepared an analysis of possible Soviet space moves. The analysis cited "[p]ublic comments in 1982 by the Soviet S[cience] & T[echnology] Attaché assigned to Washington and in 1984 by the President of the Soviet Academy of Science ," which suggested that "the Soviets have confidence in their ability to conduct such a mission." The CIA then predicted that " . . . they will choose a one-year flyby as their first step ."[26]

The analysis cited current and future indicators pointing to Soviet piloted Mars exploration. These included continuing work on a heavy-lift rocket "capable of placing into low-earth orbit about five times the payload of the present largest Soviet space launc h vehicle, thereby significantly reducing the number of launch vehicles required."[27] The "strongest current indicator," however, was "the long-duration sta ys in space by cosmonauts" aboard Salyut space stations . Potential future indicators included "a cosmonaut stay in low-earth orbit of one year duration . . . [and] space tests of a nuc lear propulsion system . . . ."[28] The CIA guessed that the first Soviet Mars flyby might occur as early as 1992, in time for the 500th anniversary of Columbus's arrival in the Americas.[29]

The JSC flyby plan for countering this possible Soviet move was prepared by Barney Roberts, who performed lunar base studies in the JSC Engineering Directorate[30] Roberts' year-long Mars flyby mission would begin with orbital assembly at the Space Station. Shuttles would deliver two expendable strap-on propellant tanks and an 18-ton Mission Module to the Station. The latter would dock with a 6-ton Command Module (not to be confused with the Apollo CM) and two 5.75-ton O TVs assumed to be in space already as part of the Space Station Program. Shuttle-derived heavy-lift rockets would then deliver 221 tons of liquid hydrogen/liquid oxygen propellants. The propellants would be loaded into the strap-on and OTV tanks just prior to departure for Mars. Spacecraft weight at Earth-orbit departure would come to 358 tons.

At the proper time, the OTV engines would ignite and burn for about one hour to put the flyby craft on course for Mars. This would empty the strap-on tanks , but Roberts advised retaining them to provide additional meteoroid and radiation shielding for the crew modules. After a six-month Earth-Mars transfer, the flyby spacecraft would spend two and one-half hours within about 20,000 miles of Mars . Closest approach would occur 160 nautical miles above the Martian surface with the flyby craft moving at 5 miles per second.

As Earth grew from a bright star to a distant disk, the astronauts would discard the strap-on tanks, then undock one OTV and redock it to the Command Module. They would enter the Command Module and discard the Mission Module and the second O TV. The OTV/ Command Module combination would slow to a manageable reentry speed using the OTV's engines, aerobrake to Earth-orbital speed, then dock at the Space Station.

Roberts found (as had planners in the 1960s) that Earth return was the most problematical phase of the flyby mission because the O TV would hit Earth' s atmosphere at 55,000 feet per second, producing fric-

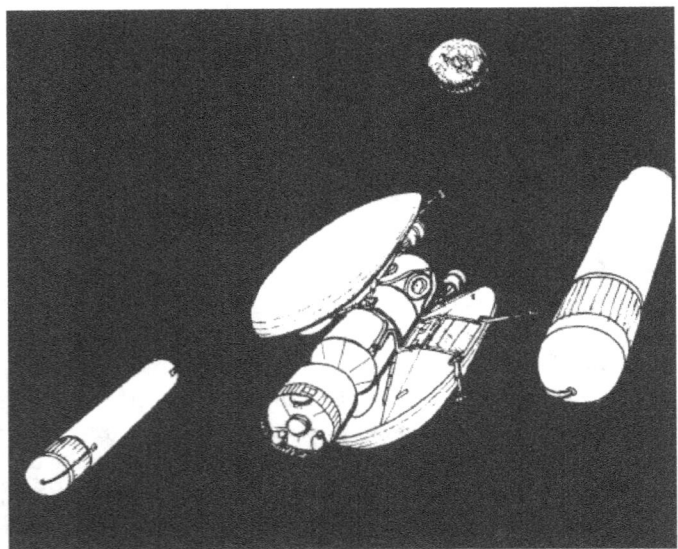

Figure 19—During return to Earth the fl yby spacecraft discards empty propellant tanks , revealing cylindrical Command and Mission Modules between twin almond-shaped Orbital Transfer Vehicles. ("Concept for a Manned Mars Flyby," Barney Roberts, Manned Mars Missions: Working Group Papers, NASA M002, NASA/Los Alamos National Laboratories, Huntsville, Alabama/Los Alamos, New Mexico, June 1986, Vol. 1, p. 210.)

tion heating beyond the planned limits of the OTV heat shields. In addition, the crew would experience "exorbitant" deceleration levels after spending a year in weightlessness.

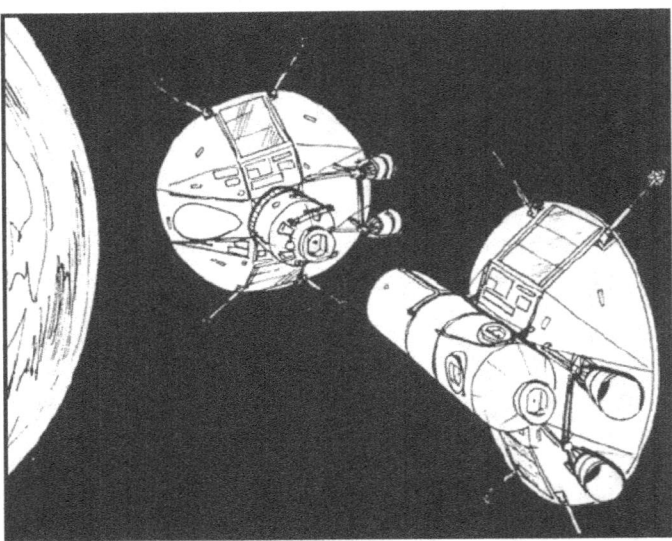

Figure 20—The flyby crew prepares to aerobrake in Earth' s atmosphere. As Earth grows from a bright star to a disk, they undock the Command Module and one Orbital Transfer Vehicle, abandoning the second Orbital Transfer Vehicle and their home for the previous year, the Mission Module. ("Concept for a Manned Mars Flyby," Barney Roberts, Manned Mars Missions: Working Group P apers, NASA M002, NASA/Los Alamos National Laboratories, Huntsville, Alabama/Los Alamos, New Mexico, June 1986, Vol. 1, p. 213.)

In the 1960s, planners proposed a Venus flyby to reduce reentry speed without using propellant, but Roberts did not mention this possibility. He proposed instead to slow the OTV and Command Module to 35,000 feet per second using the former' s engines. Adding this burn would nearly double spacecraft weight at Earth-orbit departure. Roberts calculated that, assuming the Space Shuttle-derived heavy-lift rocket could deliver cargo to Earth orbit at a cost of $500 per pound, Earth-braking propellant would add $250 million to mission costs.[31]

## Interplanetary Infrastructure

Some Mars planners envis ioned the NASA Space Station in low-Earth orbit as merely the first in a series of stations in logical places serving as Mars transportation infrastructure, much like trails, canals, railways, and coaling stations formed transportation infrastructure in bygone days. They looked ahead to solar-orbiting space stations, known as cyclers, traveling a regular path between Earth and Mars, and to spaceports at the Lagrange gravitational equilibrium points. Apollo 11 Lunar Module Pilot Buzz Aldrin, the second man on the Moon, described cyclers in a popular-audience article in Air & Space Smithsonian magazine in 1989:

> Like an oceanliner on a regular trade route, the Cycler would glide perpetually along its beautifully predictable orbit, arriving and departing with clock-like regularity. By plying the solar system's gravitational "trade winds" it will carry mankind on the next great age of exploration . . . . For roughly the same cost as getting humans safely to Mars via conventional expendable rocketry (because the problems to be solved would be largely the same), the Cycler system would provide a reusable infrastructure for travel between Earth and Mars far into the future.[32]

In the 1960s, Massachusetts Institute of Technology professor Walter Hollister and others studied "periodic" orbits related to Crocco flyby orbits but indefinitely repeating. A space station in such an orbit would cycle "forever" between Earth and the target planet. In January 1971, Hollister and his student, Charles Rall, described an Earth-Mars transport system in which at least four cycling periodic-orbit stations would operate simultaneously, permitting opportunities every 26 months for 6-month transfers between Earth and Mars.[33]

As the large periodic-orbit station flew past Earth or Mars, small "rendezvous shuttle vehicles" would race out to meet it and drop off crews and supplies for the interplanetary transfer. After several Mars voyages, the cycler approach would yield a dramatic reduction in spacecraft mass over the MOR mission mode because the cycler would only need to burn propellants to leave Earth orbit once; after that, only the small shuttles would need to burn propellants to speed up and slow down at Earth and Mars.

The Case for Mars II conference (10-14 July 1984) included a workshop that planned "a permanent Mars research base using year 2000 technology" as a "precursor to eventual colonization." The Case for Mars II workshop took advantage of the long-term weight-minimization inherent in cyclers and Mars ISRU. The

Boulder Center for Science and Policy published a JPL-funded report on the workshop results in April 1986.[34]

The Case for Mars had begun to gather steam. Participants in the second conference included Harrison Schmitt with a paper on his Mars 2000 project, Benton Clark, and Christopher McKay, who had earned his Ph.D. and gone to work at NASA Ames. Former NASA Administrator Tom Paine presented a timeline of Mars exploration spanning 1985 to 2085. It predicted, among other things, a lunar population in the thousands in the 2025-2035 decade and a Martian population of 50,000 in the 2055-2065 decade.[35] Barney Roberts, Michael Duke, and lunar scientist Wendell Mendell presented a paper called "Lunar Base: A Stepping Stone to Mars,"[36] while NASA Space Station manager Humboldt Mandell presented a paper called "Space Station: The First Step."[37]

In the Case for Mars II plan, the cycler's Earth-Mars leg lasted six months, followed by a Mars-Earth leg lasting 20 to 30 months. Each crew would spend two years on Mars, and new crews would leave Earth every two years. The first crew would leave Earth in 2007 and return in 2012; the second crew would depart in 2009 and return in 2014; and so on. This schedule would require at least two cyclers. As Hollister and Rall proposed, small Crew Shuttle vehicles would transfer crews to and from the passing cycler. The Crew Shuttles were envisioned as two-stage biconic vehicles designed for aerobraking at Earth and Mars. Their proposed shape was derived from ballistic missile warhead research.

A heavy-lift rocket capable of launching at least 75 metric tons, possibly based on Shuttle hardware, would place cycler components, some based on Shuttle and Space Station hardware, into Earth orbit for assembly. The 1984 Case for Mars plan called for cycler assembly at the low-Earth orbit Space Station; in a subsequent version, a dedicated assembly facility was proposed. The first Mars expedition would require 24 heavy-lift rocket launches and 20 Shuttle launches.

ISRU would supply the Case for Mars II base with many consumables, including propellant. "Mars is abundantly endowed with all the resources necessary to sustain life," the report stated, adding that "propellant production on the surface of Mars is critical to reducing the cost of the program" because it "reduces the Earth launch weight by almost an order of magnitude."[38] Each Crew Shuttle would require 150 tons of Mars ISRU-manufactured carbon monoxide/oxygen propellant to catch up with the passing Earth-bound cycler. The Case for Mars II workshop proposed that an automated probe should test ISRU propellant production on Mars before the Mars base program began.

## Lagrangia

Like the cycler concepts, the notion of siting infrastructure at the Lagrange points dates to the 1960s. Its theoretical roots, however, date to 1772. In that year, French mathematician Joseph Lagrange noted that gravitational equilibrium points exist in isolated two-body systems.

Lagrange points exist in space—for example, in the two-body Earth-Moon system. In theory, an object placed at one of these points will remain as if nesting in a little cup of space-time. In practice, Lagrange points in space are unstable or quasi-stable because planets and moons do not exist as isolated two-body systems. In the case of the Earth-Moon system, the Sun's gravitational pull introduces instability. Objects placed at the Earth-Moon Lagrange points thus tend to move in "halo orbits" around the Lagrange point and require modest station keeping to avoid eventual ejection.

Robert Farquhar, an engineer at NASA's Goddard Space Flight Center in Greenbelt, Maryland, first wrote about using the Lagrange equilibrium points of the Earth-Moon system in the late 1960s.[39] For the NASA MMM workshop in June 1985, Farquhar teamed up with David Dunham of Computer Sciences Corporation to propose using Lagrange points as "stepping stones" for Mars exploration.[40]

Farquhar and Dunham envisioned a large, reusable Interplanetary Shuttle Vehicle (ISV) in halo orbit about the quasi-stable Earth-Sun Lagrange 1 point, 1.5 million kilometers in toward the Sun. A Mars transport spacecraft parked there would be gravitationally bound to Earth much more weakly than if parked in low-Earth orbit. A mere propulsive burp would suffice to nudge the ISV out of halo orbit; then gravity-assist swingbys of the Moon and Earth would place it on course for Mars with little additional propellant expenditure. This meant, of course, that the amount of propellant that would need to be launched from Earth was minimized. To save even

more propellant, the ISV might park at the Mars-Sun Lagrange 1 point, about 1 million kilometers Sunward from Mars, and send the crew to the Martian surface using small shuttle vehicles.

Farquhar and Dunham pointed out that an automated spacecraft had already left Earth-Sun Lagrange 1 on an interplanetary trajectory. The International Sun-Earth Explorer-3 spacecraft had entered Earth-Sun Lagrange 1 halo orbit on 20 November 1978. After completing its primary mission it was maneuvered during 1984 through a series of Earth and Moon swingbys to place it on course for Comet Giacobinni-Zinner. Farquhar supervised the effort. The maneuvers consumed less than 75 kilograms of propellant. Renamed the International Comet Explorer, the spacecraft successfully flew past Giacobinni-Zinner, 73 million kilometers from Earth, on 11 September 1985.

Paul Keaton elaborated on Farquhar and Dunham's MMM paper in a "tutorial" paper published in August 1985. He wrote that "[a]n evolutionary manned space program will put outposts along routes with economic, scientific, and political importance" to serve as "'filling stations' for [making and] storing rocket fuel" and "transportation depots for connecting with flights to other destinations."[41]

The first outpost would, of course, be NASA's planned Space Station in low-Earth orbit, where Earth's magnetic field would help protect travelers from galactic cosmic rays and solar flare radiation, and medical researchers would learn about the effects of long-term weightlessness on the human body. Keaton then looked beyond Earth orbit for the next outpost site. He proposed placing it in halo orbit around the Earth-Moon Lagrange 2 point, 64,500 kilometers behind the Moon, or at Farquhar's Earth-Sun Lagrange 1 site. He wrote,

> [f]or the settlement of space, a Lagrange equilibrium point between the Sun and Earth has the nearly ideal physical characteristics of a transportation depot. . . . Lagrange point halo orbits are the present standard by which any alternative concept for a transportation depot must be gauged.[42]

Cyclers and Lagrange point spaceports—infrastructure spanning worlds—imply large-scale permanent space operations and long-term commitment to building space civilization. Such grandiose visions are not widely shared outside of a subset of the small community of would-be Mars explorers, as will be seen in the next chapter.

# Chapter 8: Challengers

Mars is the world next door, the nearest planet on which human explorers could safely land. Although it is sometimes as w arm as a New England October, Mars is a chilly place, so cold that some of its thin carbon dioxide atmosphere freezes out at the winter pole. There are pink skies, fields of boulders , sand dunes, vast extinct volcanoes that dw arf anything on Earth, a great canyon that would cross most of the United States, sandstorms that sometimes reach half the speed of sound . . . hundreds of ancient river valleys . . . and many other mysteries. (The Mars Declaration, 1987)[1]

## National Commission on Space

Late 1984, when the Space Shuttle was operational and Space Station development was underway, seemed an auspicious time to begin charting a course for NASA to follow after Space Station completion in the early 1990s. Congress mandated that President Reagan create an independent commission to sort through the possibilities and provide recommendations . The National Commission on Space (NCOS) w as launched officially on 29 March 1985 with the goal of blueprinting the next 20 years of the civilian space program. The NCOS was to present results to the White House and Congress following a one-year study.

Reagan tapped Tom Paine, NASA Administrator from 1968 to 1970, to head the NCOS. Fourteen commissioners joined Paine. They included such luminaries as Neil Armstrong, the first human to walk on the Moon; Chuck Yeager, the first hu man to break the sound barrier; former United Nations Ambassador Jeane Kirkpatrick; Space Shuttle astronaut Kathy Sullivan; and retired Air Force General Bernard Schriever. Laurel Wilkening, a planetary scientist and Vice Provost of the University of Arizona, was Vice Chair.

Non-voting NCOS members inc luded representatives from both parties of Congress and the Departments of State, Commerce, Agriculture, and Transportation, as well as the National Science Foundation and the White House Office of Science and Technology Policy. In addition to the inputs provided by its members , the NCOS held public forums and solicited written contributions from academe, business, and the general public.

The result was Pioneering the Space Frontier, a glossy report billed as "an exciting vision of our next fifty years in space."[2] It was the first in a series of high-profile space reports produced in the Reagan/Bush years .

Paine's attitude had not c hanged much since his time as NASA Administrator. He still sa w it as his job to challenge Americans to take on the solar system. The NCOS report's expansive vision bore Paine's unmistakable stamp; in fact, it bore a resemblance to P aine's timeline from the Case for Mars II. Paine looked to an expanding 21st-century economy with "free societies on new worlds" and "American leadership on the new frontier."

Events caught up with the NCOS exercise, however. On the chilly Florida morning of 28 J anuary 1986, with much of the Commission' s work complete , Space Shuttle Challenger exploded 73 seconds into mission STS-51L, killing seven astronauts and grounding the remaining three Shuttle orbiters. The immediate cause of the accident w as failure of a seal in one of the Shuttle's twin SRBs.

The Challenger accident threw the giddy optimism of Paine's NCOS report into sharp relief. It was a wake-up call. The Space Shuttle would not, could not, provide the kind of low-cost, routine space access envisioned during the 1970s. "The myth of an economic Shuttle" was laid bare.[3] The basic tool for establishing space infrastructure was found wanting, forcing many of the infrastructure elements envisioned by Mars planners in the early 1980s into some indefinite post-Shuttle future.

The accident contributed to NASA's decision to redesign the Space Station in mid-1986. After more than two years of studies, NASA had unveiled its station design in early 1986. Called the Dual Keel, it was primarily a space laboratory, but included a large rectangular truss whic h might eventually hold hangars, assembly equipment, and a propellant depot for Moon and Mars spacecraft. The rectangular truss was, however, adopted primarily to provide attachment points for anticipated user pa yloads, with space-facing payloads on the top and Earth-facing payloads on the bottom.[4]

Following Challenger, the Dual Keel design came to be seen as too ambitious . The rectangular truss w as deferred to a future Phase II of station assembly. Phase I would consist of a single straight truss holding solar

# Chapter 8: Challengers

arrays and a cluster of pressurized modules. Designers sought, however, to include software "scars" and hardware "hooks" in the Phase I design to permit eventual expansion to the full Dual Keel configuration.[5]

From the Mars explorers' point of view , the accident demonstrated that the Space Shuttle could not be used to launch Mars ships. It had been felt by many before Challenger that the Shuttle would ha ve to be supplemented by a hea vy-lift rocket if piloted flight beyond low-Earth orbit was to be a credible NASA goal, but it became patently obvious to most everyone on that cold day in January 1986.

Paine's report w as crammed full of new vehic les and interplanetary infrastructure based largely on SAIC and Eagle Engineering studies . The NCOS called for new cargo and passenger launch vehicles to replace the Space Shuttle by 1999 and 2000, respectively. These were components of a "Highway to Space" that would include the initial Earth-orbital Space Station (1992), a space-based OTV (1998), and an initial Earth-orbital spaceport (1998). This segued into a "Bridge Between Worlds" that would inc lude a single-stage-to-orbit space plane, a Moon base with facilities for mining lunar oxygen, cyclers and Lagrange point stations , nuclear-electric space freighters, and, by 2026, a Mars base resembling the one put forw ard at the Case for Mars II workshop (1984).

The NCOS program would cost about $700 billion between 1995 and 2020. This cost would, Paine wrote, be paid through increases in NASA funding keeping pace with projected increases in U .S. GNP of 2.4 percent per year. NASA funding in 1986 was about $10 billion, or less than 1 percent of GNP . According to the report, if NASA funding remained near 1 percent of GNP, it would increase to $20 billion in 2000 and to $35 billion in 2020. For the near-term, the report urged that the new technology development share of NASA's budget be raised from 2 percent to 6 percent.

The NCOS turned over its report to the Reagan White House in March 1986, two months after the Challenger accident. Paine went public with the report even before presenting it to the White House by giving a draft to Aviation Week & Space Technology magazine.[6] Unusually, the report w as also published as a trade paperback and sold in bookshops.

Paine presented the NCOS report formally to President Reagan and the Senate and House Space Committees on 22 July 1986. It urged the White House to direct the NASA Administrator to respond by 31 December 1986 with general long-range and specific short-range implementation plans. Paine summed up the NCOS report the next day at the NASA Mars Conference, underway at the National Academy of Sciences to commemorate the tenth anniversary of Viking 1's landing. He told the assembled scientists and engineers that Reagan had assured him that the Commission' s recommendations would be accepted.[7]

The report's conclusion assumed—correctly—that Paine's vision would be seen as grandiose, and took pains to defend it. As he had done in the 1969 Space Task Group report, Paine described the technological progress made in the past in an effort to demonstrate the progress that could be made in coming decades.

> Is our expansive view of America's future realistic? Are the tec hnical advances we project achievable? Will people accept the risks and discomforts to work on other worlds? We believe that the answer to all three questions is "Yes!" Few Americans in the early days of the Air Age ever expected to fly the Atlantic. . . yet nearly 75,000 people now fly the Atlantic daily . . . . It is equally difficult for Americans this early in the Space Age to visualize the 21st-century technologies that will enable the a verage citizen to soar into orbit at low cost, to fly to new worlds beyond Earth, and to work and live on the space fron tier in c losed-ecology biospheres using robotically-processed local resources . . . . We should . . . emphasize that: The Commission is not prophesying; it is describing what the United States can make happen through vigorous leadership in pioneering the space frontier.[8]

The NCOS plan w as not so muc h a plan for guiding NASA's future as an evocation of the pioneering spirit which Paine felt was flagging in 20th-century Americans. The romantic attraction to pioneering has in fact alw ays been a rare thing . Those afflicted by it frequently feel great zeal, which blinds them to the fact that they are rarities—that others , while frequently sympathetic to their vision, do not place as high a priority as they do upon making it real.

The NCOS report w as not well received, primarily because the Challenger accident had made clear that NASA was in no position to tackle such an expansive, all-encompassing plan. But it was also seen as too general, with too many proposals . In late August 1986, former Presidential Science Advisor George K eyworth, who had been a non-voting NCOS member, said the report had forfeited impact by putting forward proposals "that stretch all the way from China to New York."[9] At a time when NASA w as grounded and struggling to adapt its programs to the Shuttle's revealed shortcomings, the NCOS discussed topics as wide-ranging as self-replicating space factories , the International Space Year, and the Big Bang . Arguably, all were important to NASA's future missions , but presenting them in a single report merely made the view forw ard seem more clouded.

The Reagan White House quietly shelved the NCOS report; as Paine complained in an A viation Week & Space Technology opinion piece in September 1987, "[T]he mandated presidential response to the commission has been delayed."[10] It is hard to fault the spirit of Paine's report. But the Agency's challenge in 1986 w as to recover from the Challenger accident. If a plan for NASA's future in space w as to be dra wn up, it would have to attempt to take into account the realities of U.S. space flight in the mid-1980s. Such a plan was not long in coming, thanks to heightened public interest in NASA's activities following the Challenger accident, widespread concern that NASA had no long-term direction, and on-going efforts by Mars advocates .

## The Ride Report

Sally Ride was a member of the 1978 astronaut c lass, the first selected for Space Shuttle flights; in 1983 she became the first American woman in space. She flew on the Shuttle twice and sat on the Rogers Commission investigating the Challenger accident before J ames Fletcher, in his second stint as NASA Administrator, appointed her as his Special Assistant for Strategic Planning (18 August 1986) and c harged her with preparing a new blueprint for NASA's future. She was assisted by a 10-member panel and a small staff . The result of her 11-month study was a slim report entitled Leadership and America's Future in Space.

Aviation Week & Space Technology reported initial resistance inside NASA to releasing Ride' s report. The magazine quoted an unnamed NASA manager who said the agency w as "afraid of being criticized by the Office of Management and Budget." The report's frank tone may also have contributed to NASA's reluctance. In the end, Agency managers relented and published 2,000 copies in August 1987.[11]

On 22 July 1987, Ride testified to the House Subcommittee on Space Science and Applications. She told the Subcommittee that the "civilian space program faces a dilemma, aspiring toward the visions of the National Commission on Space, but faced with the realities of the Rogers Commission report."[12] Ride explained that she had attempted to reconcile "two fundamental, potentially inconsistent views." "Many people," she said, believed that "NASA should adopt a major visionary goal. They argue that this would galvanize support,focus NASA programs, and generate excitement." Others, Ride stated, maintained that NASA was "already overcommitted for the 1990s"—that it would be "struggling to operate the Space Shuttle and build the Space Station, and could not handle another major program."[13]

While Paine's NCOS report urged rapid implementation of an expansive vision, Ride's report outlined four more limited leadership initiatives "as a basis for discussion." She explained that her report w as "not intended to culminate in a selection of one initiative and elimination of the other three , but rather to provide concrete examples which could catalyze and focus the discussion of the goals and objectives of the civil space program, and of NASA efforts required to pursue them."[14]

Ride thus deviated from the pattern P aine had established in the STG report and continued in the NCOS report; she did not propose a single "master plan." In her congressional testimony she explained her guiding principle: "goals must be carefully chosen to be consistent with the national interest and . . . NASA capabilities. It is not appropriate for NASA to set the goals of the civilian space program. But NASA should lead the discussion . . . , present options, and be prepared to make recommendations."[15] Ride's four Leadership Initiatives were as follows:

- Mission to Planet Earth: "a program that would use the perspective afforded from space to study . . . our home planet on a global scale."

- Solar system exploration using robots.

- Outpost on the Moon: an ". . . evolutionary, not revolutionary . . . program that would build on. . . the legacy of the Apollo Program . . . to continue exploration, to establish a permanent scientific outpost, and to begin prospecting the Moon's resources."

- Humans to Mars: "a series of round trips to land on the surface of Mars, leading to the eventual establishment of a permanent base." The Mars mission, Ride asserted, should "not be another Apollo—a one-shot foray or a political stunt."[16]

None of Ride's four initiatives necessarily depended on the others. Her "attempt to crystallize our vision of the space program in the year 2000" in fact represented a partial break from the space station-Moon-Mars progression that had typified most NASA advanced planning.[17] Ride's approach caused confusion. For example, Aviation Week & Space Technology magazine and many newspapers incorrectly reported that she had called for a Moon base as a precursor to a piloted Mars mission. In fact, her report stated that the Moon was "not absolutely necessary" as a "stepping stone" to Mars.[18, 19]

This reflected the influence of a NASA Advisory Council Task Force led by Apollo 11 astronaut Michael Collins. "I think it is a mistake to consider the [M]oon as a necessary stepping stone to Mars," Collins told Aviation Week & Space Technology in July. "It will not get support politically, or from the U.S. public, which thinks we've 'already done the [M]oon.'"[20] Ride personally favored the Moon-Mars progression, however; she wrote that it "certainly makes sense to gain experience, expertise, and confidence near Earth first."[21]

In common with the station-Moon-Mars progression, Ride's initiatives all included NASA's Space Station. This was a ground rule established by Fletcher—not surprisingly, since the Space Station Program had begun only three years before and was fiercely defended by NASA.[22] As explained earlier, in Challenger's aftermath, Space Station had become a two-phase program. Ride pointed out that a decision on NASA's future course would impact the Phase II configuration. She wrote that a "key question for the not-too-distant future is 'how should the Space Station evolve?'" and

noted that Space Station evolution workshops in 1985 and 1986 had found that "a laboratory in space featuring long-term access to the microgravity environment might not be compatible with an operational assembly and checkout facility [of the type envisioned to support Moon and Mars exploration], as construction operations could disturb the scientific environment."[23]

Like the NCOS report, Ride's report called for NASA to increase its efforts to develop advanced space technology for exploration missions. She told the House Subcommittee that "the future of our space program lies in careful selection and dedicated pursuit of a coherent civil space strategy, and the health of our current space program lies in determined development of technologies required to implement that strategy."[24] Ride's report recommended Project Pathfinder, a program to develop technologies that had been identified by a panel of NASA engineers as crucial to future space programs. These included aerobraking, automated rendezvous and docking, and advanced chemical propulsion. "Until advanced technology programs like Pathfinder are initiated" wrote Ride, "the exciting goals of human exploration will always remain 10 to 20 years in the future."[25]

On 1 June 1987, Fletcher had created the NASA Headquarters Office of Exploration, with Ride as Acting Assistant Administrator for Exploration, responsible for coordinating missions to "expand the human presence beyond Earth." In explaining this move, Fletcher said that "[t]here are considerable—even urgent—demands for a major initiative to reenergize America's space program . . . this office is a step in responding to that demand."[26] In her report, Ride wrote that "[e]stablishment of the Office of Exploration was an important first step. Adequate support of the Office will be equally important." She noted that there was "some concern that the office was created only to placate critics, not to provide a serious focus for exploration. Studies relating to human exploration of the Moon or Mars currently command only about 0.03 percent of NASA's budget . . . this is not enough . . . ."[27]

Ride targeted the first Mars landing for 2005. Her report pointed out, however, that "NASA's available resources were strained to the limit flying nine Shuttle flights in one year." "This suggests," it concluded, "that we should . . . proceed at a more deliberate (but still aggressive) pace, and allow the first human landing to occur in 2010. This spreads the investment over a longer period."[28]

SAIC began designing the Mars mission in Ride's report in January 1987 and completed its study for the NASA Headquarters Office of Exploration in November 1987.[29] John Niehoff, the study's Principal Investigator, was the "Humans to Mars Initiative Advocate" for the Ride Report. He had also worked on The Planetary Society's 1984 Mars study (see Chapter 7). Niehoff's team proposed a three-part Mars exploration strategy:

- 1990s: Robotic missions, including a global mapper and a sample-return mission, would "address key questions about exobiology and obtain ground-truth engineering data. " This period would also see research aboard the Space Station into the effects of prolonged weightlessness on astronaut health, and development of "heavy-lift launch vehicles, high energy orbital transfer stages, and large-scale aerobrakes."

- 2000s: Piloted missions with round-trip times of about one year, stay-times near Mars of 30 to 45 days, and Mars surface excursions of 10 to 20 days were the primary emphasis of the SAIC study. These missions would explore potential outpost sites and build up interplanetary flight experience. The one-year trip-time was designed to reduce crew exposure to weightlessness and radiation.

- After 2010: "A piloted base on Mars . . . a great national adventure which would require our commitment to an enduring goal and its supporting science, technology, and infrastructure for many decades."[30]

A large amount of energy would be required to get the ship to Mars and back in about a year, which in turn would demand a prohibitively large amount of propellant. With an intent to reduce the number of heavy-lift rocket launches needed to mount the expedition, SAIC adopted a split/sprint mission mode based on a design developed by students from the University of Texas and Texas A&M University. This had a one-way, automated cargo vehicle leaving Earth ahead of the piloted sprint

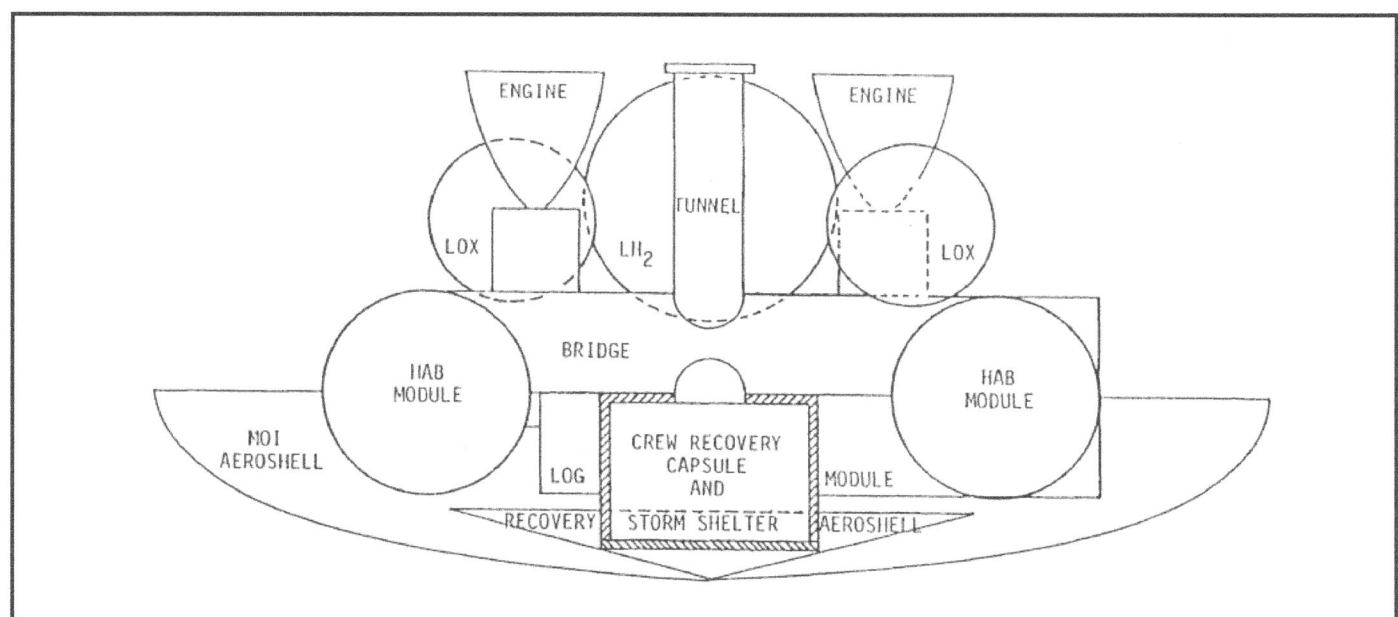

Figure 21—Science Applications International Corporation developed its Mars mission plan for NASA during 1987. The piloted spacecraft (shown here in cutaway) would reach Mars with empty propellant tanks and dock with a waiting automated cargo ship to fill up for the trip home—a controversial departure from past Mars plans . (Piloted Sprint Missions to Mars, Report No. SAIC-87/1908, Study No. 1-120-449-M26, Science Applications International Corporation, Schaumburg, Illinois, November 1987, p. 9.)

aerobrake, and transfer Earth-return propellant from the cargo vehicle. The lander crew would then return to Mars orbit in the ascent stage. On 2 August 2005, the sprint vehicle would fire its engines for a high-energy five-month sprint return to Earth.

Phase 4 would begin a few days before Earth arrival (15 January 2006 for a nominal mission). The astronauts would enter the ERV and separate from the sprint spacecraft. The ERV would aerobrake into Earth orbit while the abandoned sprint ship entered solar orbit. A station-based OTV would recover the ERV; then a Space Shuttle would return the crew to Earth.

On 26 May 1987, NASA had announced that, after finishing her study, Ride would leave NASA to become Science Fellow in the Stanford University Center for International Security and Arms Control.[33] In August, John Aaron took over the Office of Exploration. Studies begun in January 1987 to support the Ride report became the basis for piloted exploration "case studies" in FY 1988. These examined a mission to Phobos, a Mars landing mission, a lunar observatory, and a lunar outpost-to-Mars evolutionary program. All commenced with assembly of the Phase I Space Station.[34]

Martin Marietta became the Office of Exploration's de facto exploration study contractor. On 15 May 1987, NASA Marshall had awarded the $1.4-million Mars Transportation and Facility Infrastructure Study contract to the company, with SAIC in "an important teaming role," and Life Systems and Eagle Engineering as subcontractors.[35] The initial contract focus was in keeping with Marshall's propulsion emphasis; as in the EMPIRE days, the Huntsville Center anticipated developing new rockets for Mars.

However, because it was the only Mars-related NASA contract when the Office of Exploration was established, it became a mechanism for funding more general Mars-related studies. The contract, which lasted until 30 April 1990, underwent 500 percent growth as new study areas were grafted on. By the time it ended, Martin Marietta had generated nearly 3,000 pages of reports. Though Martin Marietta lost the contract to Boeing when it was recompeted in late 1989, it served to create an institutional expertise base for Martin Marietta studies during the Space Exploration Initiative (1989–93).[36]

## Opposition

NASA started as an instrument of Cold War competition with the Soviet Union. In the 1970s, having won the race to the Moon, NASA was partly reapplied as an instrument of international détente. The 1972 Space Cooperation Agreement called for the Apollo-Soyuz Test Project and other cooperative space activities. A Soviet Soyuz spacecraft docked in Earth orbit with America's last Apollo spacecraft in July 1975. When the agreement was renewed in 1977, it included plans for a U.S. Shuttle docking with a Soviet Salyut space station. By 1980, however, the Soviet invasion of Afghanistan had undermined détente, ending virtually all talks on piloted space cooperation.[37]

In 1982, the Reagan White House let the Space Cooperation Agreement lapse to protest continued Soviet involvement in Afghanistan and martial law in Poland. In the first major step toward renewed cooperation, Senator Spark Matsunaga (Democrat-Hawaii) sponsored legislation calling for renewal of the Space Cooperation Agreement. Congress passed the Matsunaga resolution, and President Reagan signed it into law in October 1984.[38]

On 11 March 1985, Mikhail Gorbachev became the Soviet Union's new leader. He set about implementing a raft of new reform policies. Making them work meant diverting resources from Cold War confrontation to domestic production. A charismatic leader representing a new generation of Soviet politicians, he encouraged many in the West by working to thaw relations with the United States.

Against this background, The Planetary Society partnered with the influential AIAA to hold the Steps to Mars conference in Washington, DC, on the tenth anniversary of Apollo-Soyuz. NASA Administrator James Beggs was on hand to hear Carl Sagan and others promote a joint United States-Soviet Mars expedition.

Sally Ride had written of the difficulty of reconciling visionary and conservative space goals. The Planetary Society Mars proposal fell into the former category. Unlike some visionary goals, however, it proposed giving Mars exploration a political purpose, just as Apollo lunar exploration had a political function in the 1960s. Beggs endorsed U.S.-Soviet space cooperation, but cau-

# Chapter 8: Challengers

tioned that "when you get down to the nitty-gritty of working out details, it's not so easy."[39]

The U.S. and the Soviet Union renegotiated a new Space Cooperation Agreement in November 1986. Unlike its predecessors in 1972 and 1977, it contained no provision for cooperative piloted missions. A month later, Sagan published a prescient editorial in Aviation Week & Space Technology. The Cornell University astronomer asked, "What if sometime in the next few years a general strategic settlement with the Soviet Union is achieved . . . ? What if the level of military procurement . . . began to decline?" Sagan believed that "[I]t [was] now feasible to initiate a systematic program of exploration and discovery on the planet Mars . . . culminating in the first human footfalls on another planet" at a cost "no greater than a major strategic weapons system, and if shared by two or more nations, still less." He added that Mars was "a human adventure of high order, able to excite and inspire the most promising young people."[40]

The U.S. and the Soviet Union renewed the Space Cooperation Agreement in April 1987. Emboldened, The Planetary Society circulated The Mars Declaration widely in late 1987. Declaration signatories included former NASA Administrators and Apollo-era officials, astronauts, Nobel laureates, actors, authors, politicians, university presidents and chancellors, professors, pundits, composers, artists, and others. It called for a joint U.S.-Soviet expedition to serve as a model for superpower cooperation in tackling problems on Earth, and it called Mars a "scientific bonanza" that could provide "a coherent focus and sense of purpose to a dispirited NASA" in the wake of the Challenger accident.[41]

Mars, the Declaration continued, would give the U.S. Space Station a "crisp and unambiguous purpose" as an assembly point for Mars ships and as a laboratory for research into long-duration space flight. Planetary Society vice president and former JPL director Bruce Murray was outspoken on this point. Reiterating what George Low and Nixon's PSAC had stated in the early 1970s, he told the AIAA in January 1988 that "the principal logic for the [S]tation is in the context of a Mars goal."[42]

Meanwhile, the "future indicators" the CIA had listed for Harrison Schmitt in 1985 had begun to occur. On 15 May 1987, the Soviet Union launched the first Energia

rocket, the most powerful to leave Earth since the U.S. scrapped the Saturn V. Energia functioned perfectly, though its 80-metric-ton Polyus payload failed to achieve orbit. On 21 December 1988, cosmonauts Vladimir Titov and Musa Manarov returned to Earth after a record 365-day stay aboard the Mir space station—long enough to have performed a one-year piloted Mars flyby.

Mikhail Gorbachev first publicly called for a joint U.S.-Soviet Mars mission as Titov and Manarov boarded Mir in December 1987. He told the Washington Post and Newsweek before the May 1988 Moscow summit that he would "offer to President Reagan cooperation in the organization of a joint flight to Mars. That would be worthy of the Soviet and American people."[43] On 24 May 1988, Pravda carried an article by Soviet space flight leaders Yuri Semyonov, Leonid Gorshkov, and Vladimir Glushko calling for a joint Mars mission. [44] Little progress was made toward Mars at the Moscow Summit, but major strides were taken toward ending the Cold War. Time magazine's cover for 18 July 1988 showed a Viking photo of Mars with U.S. and Soviet flags and the legend "Onward to Mars."

Halfway through Titov and Manarov's year-long stay on Mir (7 July 1988), the Soviet Phobos 1 Mars probe lifted off from Baikonur Cosmodrome on a Proton rocket. Phobos 2 lifted off on 12 July. The twin probes featured involvement by more than a dozen countries, including the United States. They were designed to orbit Mars and explore their namesake moon Phobos. After rendezvous with the pockmarked little moon, they would drop a "hopper" rover and a small lander.

In retrospect, however, the probes were the Soviet Mars program in miniature—they got off to a triumphant start, then sputtered. On 31 August 1988, operators at the Flight Control Center in Kaliningrad, near Moscow, sent the Phobos 1 Mars spacecraft an erroneous radio command that caused it to lose attitude control and turn its solar arrays away from the Sun. Starved for power, Phobos 1 failed just two months into its 200-day flight to Mars. Phobos 2 reached Mars orbit on 29 January 1989. The spacecraft returned useful data on Mars and Phobos; however, it failed in late March as it neared the long-anticipated Phobos rendezvous.

At 6.5 metric tons each, the Phobos probes were the heaviest Mars probes ever to leave Earth orbit. They took advantage of the minimum-energy launc h opportunity associated with the September 1988 Mars opposition, the best since 1971.

Mars glowed bright orange-red in Earth' s skies as the Space Shuttle Discovery was rolled to its Florida launch pad for the first Shuttle flight since Challenger. On 29 September 1988, as Earth overtook Mars in its orbit and pulled ahead, Discovery lifted off on the 26th flight of the Space Shuttle Program. The four-day, five-crew STS-26 flight ended a 33-month hiatus in U .S. piloted space flight—the longest since the 1975-1981 Shuttle development period. By the time STS-27 launched in December, Mars was fading fast and the U.S. Space Shuttle w as no longer the world' s only reusable piloted spacecraft. The second Energia rocket

had launched on 15 November 1988 with a Buran shuttle on its back for an unpiloted test flight.

The Mars planning community , though still small and with few resources, was in ferment. New leadership in the Soviet Union, the expanding Soviet space program, and the tha wing of U .S.-Soviet relations, coupled with America's return to piloted space flight and growing public awareness of Mars, seemed to create an opportunity. As will be seen in the next c hapter, newly elected President George Bush would take up the mantle of President K ennedy and declare for Mars. Though a failure, his initiative would not be without significant results.

# Chapter 9: Space Exploration Initiative

The study and programmatic assessment described . . . have shown that the [Space] Exploration Initiative is indeed a feasible approach to achieving the President's goal . . . . The last half of the 20th century and the first half of the 21st century will almost certainly be remembered as the era when humans broke the bonds that bound them to Earth and set forth on a journey into space . . . . Historians will further note that the journey to expand the human presence into the solar system began in earnest on July 20, 1989, the 20th anniversary of the Apollo 11 landing. (The 90-Day Study, 1989)[1]

## Space Wraith

Viewed as a space program, as it was intended to be, the Space Exploration Initiative (SEI) was a failure. Viewed as an "idea generator" for Mars exploration planning, however, SEI was a success—some concepts it fostered dramatically reshaped subsequent planning efforts.[2] It was also successful as a painful but necessary growth process. SEI relieved NASA of weighty historical baggage. It weaned large segments of the Agency from its faith in the efficacy of Kennedyesque Presidential proclamations, and it further weakened the pull the station-Moon-Mars progression exerted on senior NASA managers, a process that had first seen high-level expression at NASA in the 1987 Ride report.

Like Apollo before it, the decision to launch SEI had more to do with non-space policy than with its stated space flight aims. SEI and Apollo were, however, diametrical opposites in most other respects. Apollo occurred at the Cold War's height, while SEI occurred at its end. Apollo aimed at displaying American technological prowess to counter Soviet space successes, while SEI aimed in part to provide new tasks for defense-oriented government agencies and contractors as the Soviet threat receded. Apollo was greeted with public enthusiasm, while SEI was forgotten even as it began. Finally, Apollo accomplished both its political and space flight goals, while SEI accomplished neither.

The concept of a big Apollo-style space initiative was in the air in the late 1980s. In late 1987 and early 1988, the Reagan Administration considered and rejected a "Kennedy-style declaration" calling for a Moon base or a man on Mars. White House staffers explained that they had lacked information adequate to make a "technically and fiscally responsible decision."[3] The White House opted instead for its National Space Policy (February 1988) and for giving NASA's Space Station a name—Freedom (July 1988). More importantly, it requested $100 million in FY 1989 to start NASA's Pathfinder technology development program. The Agency had asked for $120 million. In December 1987, a National Research Council report estimated that NASA would have to spend $1 billion a year on technology development for several years to make up for past neglect. Despite this finding, the funding request was poorly received in Congress—not a propitious sign for big new initiatives.[4]

Early in his Administration, President George Bush reestablished the National Space Council and put his Vice President, Dan Quayle, in charge. On 31 May 1989, Bush directed NASA to prepare for a Presidential decision on America's future in space by proposing a space goal with visible milestones achievable early in the 21st century. The directive was said to have originated with OMB director Richard Darman and Quayle's advisors.[5] NASA Administrator Richard Truly, Assistant Administrator for Exploration Franklin Martin, and JSC director Aaron Cohen briefed Quayle in June.

Bush revealed what they had told Quayle and launched the SEI on the steps of the National Air and Space Museum on 20 July 1989, the 20th anniversary of the Apollo 11 Moon landing. Bush told his audience,

> Space is the inescapable challenge . . . . We must commit ourselves to a future where Americans and citizens of all nations will live and work in space . . . . In 1961 it took a crisis—the space race—to speed things up. Today we don't have a crisis. We have an opportunity. To seize this opportunity, I'm not proposing a 10-year plan like Apollo. I'm proposing a long-range, continuing commitment. First, for the coming decade—for the 1990s—Space Station Freedom, our critical next step in all our space endeavors. And next—for the new century—back to the Moon. Back to the future. And this time, back to stay. And then, a journey into tomorrow, a journey to another planet—a manned mission to Mars . . . today I'm asking . . . our

# Chapter 9: Space Exploration Initiative

able Vice President, Dan Quayle, to lead the National Space Council in determining specifically what's needed . . . . The space council will report back to me as soon as possible with concrete recommendations to chart a new and continuing course to the Moon and Mars and beyond.[6]

Aviation Week & Space Technology greeted the initiative with skepticism and a pun, calling it a "space wraith." "President Bush," the magazine reported, "set forth a long-term space plan without a budget and with no more than a skeletal timetable . He then called for more study."[7] The initiative w as, moreover, "sprung" on Congress with little "spadework" by either the Administration or by NASA.[8] This helped ensure opposition. Congress sent the Bush White House a clear message by eliminating funds for the Pathfinder technology development program from the FY 1990 NASA budget.

On 26 July, Truly and Martin briefed NASA employees on the President's call. They outlined a "building block approach to progressively more difficult human missions."[9] The proposal, a retread of the 1960s Integrated Program Plan, ignored the less expansive alternative program approach laid out by Sally Ride in 1987. Ride's report was based on conservative projections of NASA's future resources, but the Truly and Martin plan took it for granted that resources for a large, Apollo-style program would automatically follow the President's Kennedyesque proclamation.[10]

Truly and Martin laid out the following timetable:

* 1995-2000: Space Station Freedom operational; robotic precursor spacecraft explore the Moon

* 2001-2010: Lunar outpost; robotic precursors explore Mars

* Post-2010: Mars expedition

The Moon and Mars goals would give "direction and focus" to Space Station F reedom, Truly and Martin stated, while the lunar outpost would give American astronauts experience in living and working on another world before confronting Mars' greater demands. The Moon's proximity to Earth ("a three-day trip") and scientific value made it an attractive way station on the road to Mars . For its part, Mars was

SEI's ultimate goal because it had "intrigued humans for centuries," was "scientifically exciting" and "the most Earth-like planet," and because it had resources to support human life. Striving for Mars would "cement long-term U.S. leadership in space" by providing a "challenging focus for [the] space program."

Truly and Martin told NASA civil servants that Bush's call was "a major institutional c hallenge for NASA" that would "require restructure of [the] [A]gency." NASA would seek to add staff and facilities and would streamline its procurement system.

## The 90-Day Study

The task of turning NASA's SEI plan into a report for Quayle's National Space Council fell to an internal NASA team led by Aaron Cohen. He had 90 da ys to complete his study, which started on 4 August 1989. The schedule was said to ha ve been driven in part by Bush's desire to have an SEI implementation plan in hand for his State of the Union speech in early 1990. In September, Truly said that Cohen's study involved 160 managers from across the Agency, of whom 100 were based at JSC . Mark Craig, JSC Lunar-Mars Exploration Program Office manager, headed the JSC study team.[11] On 2 November 1989, Truly passed Cohen's report to President Bush.

Cohen's report contained five "reference approaches" that followed "the President's strategy: First, Space Station Freedom, and next back to the Moon, and then a journey to Mars." There was, of course, nothing new to this approach. In common with the 1969 Space Task Group report, the reference approaches were in fact one approach with multiple timetables for carrying it out, not a range of alternate plans. NASA seemed to be saying that there w as only one w ay to explore the Moon and Mars.

Approach A emphasized "balance and speed." Space Station Freedom assembly would be completed in 1997, two to three years ahead of the completion date planned at the time Cohen' s report w as released. Astronauts would return to the Moon in 2001 and permanently staff a lunar outpost the following year . By 2010 the outpost would produce 60 tons of oxygen per year. In 2016, four astronauts would travel to Mars in a transfer vehicle using lunar oxygen propellant and would spend 30 da ys on the surface . The year 2018 would see the first 600-da y tour-of-duty in a permanent Mars outpost.

Cohen's Approach B aimed for the "earliest possible landing on Mars." The lunar and Mars activities outlined in Approach A would occur simultaneously , requiring more spending in the 2000-2010 decade . The first Mars expedition would occur in 2011.

Approach C strove for "reduced logistic support from Earth"—that is, increased reliance on ISRU . Lunar oxygen production would thus begin in 2005, earlier than in Approach A.

Approach D was to pick Approach A, B, or C, then slip all dates two to three years to allow Space Station Freedom completion in 1999 or 2000. Americans would return to the Moon in 2004.

Approach E assumed that the U .S. would undertake Bush's initiative, but on a "reduced scale." Freedom would be completed as sc heduled in 1999 or 2000, and Americans would return to the Moon in 2004. The lunar outpost would be completed in 2012, and astronauts would spend 30 da ys on Mars in 2016. A 60-day Mars stay would occur in 2018, followed by a 90-day stay in 2022. The Mars outpost would be activated in 2027.

Cohen's report called for new hea vy-lift rockets based on Space Shuttle hardw are or on the P entagon's Advanced Launch System. The largest would place up to 140 metric tons into orbit and ha ve a launch shroud up to 15 meters wide—large enough to cover reusable aerobrake heat shields.

To support the Bush initiative, Space Station Freedom would evolve from lab to spaceport through four configurations. First, the baseline single truss would be expanded to include the vertical lower keel trusses and lower boom truss of the Dual K eel design. The second configuration saw the addition of a lunar spacecraft hangar and a second habitatio n module to house four -person crews en route to the Moon. The crew roster would rise to 12 in the th ird configuration to support lunar spacecraft servicing and increased life sciences research and Freedom maintenance. The fourth configuration would see the addition of the Dual Keel upper trusses and installation of a Mars s pacecraft assembly facility.

The report called for increased civil service hiring and new budget processes within NASA, but it included no cost estimates . A JSC team led by Humboldt Mandell performed a cost analysis and prepared a cost section, but it was stricken and most copies shredded by Truly's order because the costs arrived at were deemed politically unacceptable .[12] Cost information w as leaked from the National Space Council, however, so suppressing the cost data merely stymied informed discussion. [13] SEI's critics seized on the highest leaked cost estimates without consideration of the cushion they contained because they lac ked complete information—or , if they had access to the details of the cost estimate , they could safely ignore them because they were not publicly available.

According to Mandell, The 90-Day Study plan "was over-costed by a considerable amount."[14] The stricken cost estimates inc luded a 55 percent reserve—"an allowance incorporating both the cost estimating uncertainties for individual developments (i.e., project-level reserves) and allow ances for c hanges in scope (i.e., program-level reserves)."[15] The initial cost of a permanent Moon base using Approach A and including the 55 percent "cushion" would be $100 billion in constant 1991 dollars between 1991 and 2001. The Mars expedition would cost an additional $158 billion between 1991 and 2016 based on the same stipulations. Thus, achieving the letter of Bush' s speech—a return to the Moon to sta y and a mission to Mars— would cost a total of $258 billion, of which 55 percent ($141 billion) was cushion.[16]

Continuing operations would, of course, add to SEI's cost. In Approach A, operating the lunar base from 2001 to 2025 would cost $208 billion, while operating a Mars outpost from 2017 to 2025 would cost $75 billion. Thus the SEI program cost for Approach A for 34 years, from 1991 to 2025, including operations and a 55 percent cushion, would come to $541 billion.[17]

The cost summary had NASA's annual budget climbing from about $13 billion in 1990 to about $35 billion in 2007 for Approach A. At its peak, about half would be allotted to Moon and Mars programs, meaning that the average annual cost for Moon and Mars would be about $15 billion per year.[18]

## A Quick Study

The 90-Day Study plan was NASA's official proposal for accomplishing SEI, but it was not the only SEI plan put forward by the Agency. In the summer of 1989, an Office of Exploration task force under Ivan Bekey performed a "quick study . . . with analysis support by Martin Marietta" which, it claimed, "defined a much more practical Mars program . . . by virtue of reducing the scale of operations through judicious choices and invention of a new launch vehicle concept."[19] Focused on Mars and relying heavily on Mars ISRU for propellant production, it appears in retrospect as a premonition of piloted Mars planning in the 1990s.

The study was based on Martin Marietta work performed under the Marshall Transportation Infrastructure contract, as well as on the Phobos and Mars Case Studies . Bekey's task force briefed Truly on its proposal in the summer of 1989, but it had little obvious influence on The 90-Day Study.[20] Bekey presented the concept at the 40th International Astronautical Federation Congress in Malaga, Spain, in October 1989, just before Truly sent Cohen's report to President Bush.

The Bekey task force proposed that astronauts go first to Phobos. There they would set up an ISRU propellant plant for making propellants from Phobos materials , which are believed to be water-rich. Bekey's group also proposed to minimize impact on Space Station Freedom by using heavy-lift launch vehicles to launch a few large components rather than resorting to on-orbit assembly of many small components . Mission rate would be kept low to reduce spending rate . The piloted Mars expedition would be preceded in the 1990s by a "preparatory program" including automated precursors, technology development, and biomedical research. The Moon played no mandatory role in Bekey's proposed Mars program.

Bekey's task force found that, assuming an opposition-class trajectory with a Venus flyby for the initial Phobos expedition and a conjunction-class trajectory for the Mars landing expeditions, the maximum spacecraft mass at Earth-orbit departure for a Phobos expedition was similar to the minimum mass for a

Mars landing expedition—about 700 tons. Therefore, the short Phobos mission in 2004 could act as a "shakedown cruise" for the Mars landing mission spacecraft design, helping to minimize risk to the crew during the longer landing missions.

Three astronauts would travel to Phobos with a piloted Mars lander , which would touch down unpiloted on Mars to act as a backup habitat for the 2007 Mars landing expedition crew . The 2004 crew would spend a month at Phobos , during which they would demonstrate an automated ISRU pilot plant.

Three expeditions would then travel to Mars' surface to set up infrastructure for a Mars outpost. Five astronauts would launch to Mars in 2007, land near the backup habitat from the 2004 mission, and spend a year on the surface. On the next expedition, five astronauts would set up the first half of a propellant production facility on Phobos and land on Mars . Expedition 4 would set up the remainder of the propellant plant, "readying the Mars infrastructure for a sustained series of visits" that would establish a permanent outpost on Mars.

The task force Phobos/Mars spacecraft design consisted of a large dish-shaped aerobrake with twin Space Station Freedom-derived cylindrical habitats . The spacecraft would rely on tethers to create artificial gravity; the astronauts would reel out the habitat modules from the aerobrake, then rotate the assemblage end over end to produce artificial gravity.

Bekey proposed launching Mars ship components and propellant on Shuttle-derived Shuttle-Z rockets, which would include three or four Space Shuttle Main Engines (SSMEs), a strengthened External Tank, and two Solid Rocket Boosters. Shuttle-Z would use existing Kennedy Space Center Shuttle facilities and cost about the same per launch as the Shuttle, "but with 4-6 times the payload."[28]

By using the Mars transfer stage as the Shuttle-Z third stage, up to 164 tons could be placed in low-Earth orbit. This would permit the Phobos/Mars spacecraft to be fully assembled with "at most" two Shuttle-Z launches. Three Shuttle-Z launches would refuel the Mars transfer stage in orbit. A similar concept was proposed in the 1971 MSC PMRG study. The crew would then board the ship from a Space Shuttle orbiter and fire the refueled transfer stage to leave Earth orbit for Mars.

The Bekey task force estimated that the total weight launched per year to carry out its Mars program would be about half that needed to carry out the split-sprint mission plan defined by SAIC for the 1987 Ride Report. Bekey's admittedly optimistic preliminary cost estimate was $40 billion for two landings on Mars.[20]

## The Great Exploration

Alternatives to The 90-Day Study also surfaced outside NASA. In mid-September, at about the time Cohen presented his initial briefing on his study to the National Space Council, Lawrence Livermore National Laboratory (LLNL) engineers, led by Lowell Wood, briefed Quayle on their Great Exploration plan for SEI.[21] LLNL, which was operated by the University of California under contract to the U.S. Department of Energy, was associated with design and test of nuclear weapons, as well as research into advanced particle beam and laser weapon systems.

The Livermore plan was not well received by NASA, which saw it as an effort to invade its territory.[22] The meaning of the "opportunity" Bush mentioned in his 20 July speech thus seemed clear—SEI was to be an opportunity for the national laboratories to expand their bailiwick. According to some participants, one purpose of SEI was to provide work for Federal government agencies and contractors suffering cut-backs because the Cold War was ending. Cohen's study had, in fact, taken into account the need to provide tasks for organizations such as the Army Corps of Engineers and the Department of Energy labs.[23] NASA's understanding was, however, that NASA would be in charge.[24]

LLNL's Great Exploration plan drew on its 1985 Columbus lunar and 1988 Olympia Mars studies.[25, 26] Wood and his colleagues explained that their plan respected "contemporary politico-economic realities," which would not tolerate a $400-billion space program lasting three decades. Their plan, they claimed, would require a decade and cost only $40 billion.[27]

The Livermore team called for "manned space exploration as though it were a profit-seeking enterprise" with "swift exploration, settlement and infrastructure

creation." "Each step," they explained, would leave "major operational legacies—and commitments" so that "Lunar and Martian Bases, once manned, never need be unmanned thereafter." They also called for extensive use of off-the-shelf technology to launch and outfit inflatable structures ("community-sized space suits"), including an Earth-orbital station, "Gas Station" propellant depots, and Moon and Mars surface bases.[28]

The Great Exploration program would commence in mid-1992, when a single Titan VI or HL Delta rocket would launch a 50-metric-ton folded Earth Station and Gas Station payload with an Apollo CM on top. The stations would deploy and inflate automatically in orbit under the crew's supervision. The Earth Station would consist of seven 15-meter-long sausage-shaped modules arranged end to end. It would rotate end over end four times each minute to create artificial gravity that would vary from deck to deck over the length of the station, thus providing crews with lunar and Martian gravity experience. The Gas Station would use solar power to electrolyze water into liquid hydrogen/liquid oxygen spacecraft propellants. Water would be launched by competing companies and purchased by the government from the lowest bidder.

In late 1994, a single rocket would launch a 70-metric-ton folded Lunar Base with an Apollo CM-based Earth Return Module on top. The Lunar Base would refuel at the Gas Station, fly to the Moon, and inflate on the surface. The astronauts would live in Spartan conditions, with crew rotation every 18 months. A lunar surface fuel factory and lunar-orbit Gas Station would be established when the second crew arrived in late 1996.

The 70-metric-ton Mars Expedition ship would be launched in late 1996, inflated in Earth orbit, and refueled at the Gas Station. It would then fly to Mars orbit and visit Phobos or Deimos before landing on Mars. The Mars Base would inflate on the surface, and the first crew would move in for a 399-day stay. They would mine Martian water to manufacture propellants for a rocket-powered hopper.

The plan was innovative, but could it work? NASA managers and engineers thought not. The national laboratories, however, had supporters in the White House and on the National Space Council, among them Vice

President Quayle. They held up the LLNL proposal as a good example of "innovative thinking."[29]

Faced with two rival plans for carrying out his initiative, in December Bush asked the National Research Council (NRC) to examine the studies . H. Guyford Stever, science advisor to Presidents Nixon and F ord and a former Director of the National Science Foundation, chaired the NRC's Committee on Human Exploration of Space. Among its 14 members were Apollo 10 astronaut Thomas Stafford and (until his death on 31 January 1990) Apollo program manager Samuel Phillips.

The Stever Committee report, unveiled on 7 March 1990, stated that the LLNL approach entailed "relatively high risk" and underestimated "the many practical and difficult engineering and operational challenges" of exploring space .[30] The report threw cold water on the push to give SEI over to the national laboratories by stating that "NASA has the organizational expertise and demonstrated capability to conduct human space exploration . . . . To attempt to replicate such expertise elsewhere would be costly and time-consuming."[31]

The Stever Committee also pointed out a basic truth applicable to all large space projects: that the "pace at which the initiative should proceed, while clearly influenced by scientific and technical considerations, is inherently determined by social and political decision-making processes in whic h non-technical constraints, such as the sustainable level of resource commitment and acceptable level of risk[,] are paramount. "[32] In other words, policy makers bore as much responsibility for setting SEI's pace, price tag, and chances for eventual success as the engineers, and they would have to make firm decisions before the engineers could plan effectively and proceed.

The Stever Committee then called for more studies , stating that "the [N]ation is at a very early stage in the development" of its Moon and Mars plans (this despite the many studies performed inside and outside NASA over the decades). "None of the analyses to date—The 90-Day Study, The Great Exploration, or, indeed, this report—should be regarded as providing more than a framework for further discussion, innovation, and debate,"[33] it stated, then added that " . . . the eventual choice of mission architecture will incorporate the ideas from a variety of concepts, some

that now exist and others that will arise in the future . . . . The variety of concepts should be regarded as a 'menu' of opportunities."[34]

In late February, a week before the Stever Committee report was publicly released, President Bush directed that NASA should be the "principal implementary agency" for SEI, with the Departments of Defense and Energy in "major roles."[35] Within a week of the report's release, President Bush followed its advice and called for more study. He asked that at least two substantially different reference architectures for SEI be produced over the next several years.

Idea collection for the Stever Committee's "menu" had begun in mid-J anuary 1990, when the Aerospace Industries Association, an organization representing aerospace contractors, had met to start a process of gathering ideas to turn over to NASA. The Agency's Office of Aeronautics and Space Technology served as ad hoc coordinator for this effort.[36] NASA also enlisted Rand Corporation to manage a campaign to solicit ideas from industry, universities, national labs, and the general public. NASA Administrator Truly led a U .S. Government interagency effort. This broad gathering of ideas became known as the SEI Outreac h Program.

Ideas collected through the Outreach Program were to be reviewed by an independent SEI Synthesis Group , which would then issue a report. The Synthesis Group approach had been recommended by the Aerospace Industries Association in April. On 16 May 1990, Congress agreed to provide $4.55 million for the Outreach Program, but not without a price. NASA had to agree that it would release no SEI-related contracts to industry until 1991. As one congressional staffer explained, this deferment was designed "to avoid raising expectations in the private sector, given the incredible [Federal] budget restraints ." The Agency also agreed to defer $5 million in internal NASA study work until 5 August 1990.[37] On 31 May, Truly introduced Tom Stafford as Synthesis Group chair.

Paul Bialla, NASA Programs Manager for General Dynamics, expressed well the skepticism many in industry felt tow ard the SEI Outreac h Program. "For the most part, our ideas have already been shared with NASA," he told Space News. "Throwing the door open to everyone is simply going to delay the process."[38]

## A Political Liability

The Outreach Program was SEI's most far-reaching contribution to Mars expedition planning , for it compiled a large body of ideas for how to send humans to Mars . In terms of implementing SEI, however, the Outreach Program amounted to a means of allowing the abortive initiative to fade quietly after it had become an obvious political liability for the Bush Administration.

Even as the Outreach Program began, SEI was mortally wounded. The Bush Administration's NASA budget request for FY 1991 was $15.1 billion, a 23 percent increase over FY 1990. This included $216 million to start SEI. Two days of NASA budget hearings in mid-March 1990 showed, however, that the Moon and Mars initiative enjoyed almost no support in Congress. By the summer of 1990, it was writ large—no matter what good ideas the Outreac h Program might produce, SEI stood almost no chance of gaining congressional support.

It was a two-part problem. On the one hand, the Democrat-controlled Congress w as not eager to hand the Republican Bush Administration any victories , especially after it had cast its 1988 Presidential candidate, Michael Dukakis, as a spend-thrift Democrat. [39] More importantly, however, the late 1980s and early 1990s were marked by an enormous F ederal debt—$3 trillion in 1990—and annual budget deficits . Budget problems alone made it unlikely that a new space initiative would be well received, even if it didn't ha ve a rumored price tag of half a trillion dollars .

On fiscal grounds, SEI opposition was bipartisan. Bill Green (Republican-New York), a member of the House Appropriations Committee, said, "[G]iven the current budget situation, I would not anticipate a significant start on Mars in the near future ."[40] Robert Traxler (Democrat-Michigan), chair of the House Subcommittee on Housing and Urban Development and Independent Agencies, summed it up succinctly: "Basically, we don't have the money."[41]

On 1 May 1990, President Bush called congressional leaders to the White House to lobby for SEI. Richard Darman sought to dec lare NASA's proposed budget increase exempt from mandatory cuts imposed by Gramm-Rudman deficit reduction legislation, and

Bush proposed that aerospace tec hnology cuts should come from the Defense budget, not from NASA. [42] The congressional response was quick in coming. On 3 May 1990, Senator Albert Gore (Democrat-Tennessee), chair of the NASA Authorization Panel, told his fellow legislators that "before discussing a mission to Mars , the Administration needs a mission to reality."[43]

Bush used his 11 May commencement address at Texas A&M University to signal SEI' s importance to his administration. His speech was historic—in it he became the first U.S. President to set a target date for an American expedition to Mars . "I am pleased to announce a new age of exploration," he told the crowd, "with not only a goal but also a timetable: I believe that before America celebrates the 50th anniversary of its landing on the Moon [in 2019], the American flag should be planted on Mars."[44]

Congress, however, handed Bush his first c lear defeat in mid-June, when a House panel eliminated all funds for SEI from the FY 1991 NASA budget. On 20 J une Bush declared that he would fight for his Moon and Mars program. His Administration had, he said, "matched rhetoric with resources." The full House eliminated all SEI funds at the end of J une.[45]

On top of issues of party and finance were badly timed NASA problems not directly related to SEI. These raised inevitable questions about the desirability of committing the Agency to a major new initiative when it appeared it could not handle what it already had. In late June, NASA announced that the $1.5-billion Hubble Space Telescope, launched into orbit on 24 April 1990, was rendered myopic by an improperly manufactured mirror. At the same time , the Shuttle fleet w as grounded by persistent hydrogen fuel leaks . The three orbiters sat on the ground for five months while NASA engineers struggled with the problem.

On the first anniversary of Bush's SEI speech, a NASA panel headed by former astronaut and spacew alker William Fisher announced that Space Station Freedom would need 6,200 hours of maintenance spacew alks before it w as permanently staffed and 3,700 hours of maintenance spacewalks each year thereafter . This would cut deeply into time a vailable for researc h aboard the orbiting space laboratory .[46] The Fisher Panel's findings helped lead to a new round of station redesign in 1990 and 1991. In an effort to reduce cost and complexity, the potential for Phase II expansion to the Dual Keel design was eliminated, along with the option for hangars, fueling facilities, and other Moon- and Mars-related systems.[47]

Space Station Freedom thus lost virtually all hope of being useful for Mars transportation. It remained important, however, as a place to gather data on the biomedical effects of long-duration space flight as part of efforts to minimize risk to future Mars crews . Not coincidentally, Mars plans that ignored the Station, except to say that they did not intend to use it, began to proliferate. NASA internal planning, however, con- tinued to place Space Station Freedom—or some future station—squarely on the path to Mars.

In October, House and Senate conferees agreed to an FY 1991 NASA budget of $13.9 billion. While this con- stituted an increase of $1.8 billion over NASA 's FY 1990 budget, it included no funds for SEI. Bush bowed to the inevitable and signed the appropriation into law.

## A New Initiative

By the fall of 1990, the course of piloted space flight over the next decade was taking shape. Bush had men- tioned international space cooperation in his speech of 20 July 1989. SEI, however, stressed U.S. space leader- ship, which implied competition with the Soviet Union. The Soviets had built up an impressive space infra- structure in the 1970s and 1980s . By 1990, however, with economic and political reforms underway in their country, they could no longer afford to use it.

As early as March 1990, Bush had directed the National Space Council to pursue space cooperation with the Soviets in an effort to encourage and support Mikhail Gorbachev's on-going reforms. On 8 July 1990, Bush agreed to let U. S. commercial satellites fly on Soviet rockets. On 25 July 1990, the United States and Soviet Union agreed to fly a NASA Mission to Planet Earth instrument on a Soviet satellite scheduled for launch in 1991. In October 1990, Quayle told reporters that "we are in serious discussions with the Soviet Union" on flying an American astronaut on Mir and a Soviet cosmonaut on the Shuttle.[48]

Yuri Semyonov, director of NPO Energia, the leading Soviet astronautics design bureau, promoted joint U.S.-

Soviet piloted Mars exploration at space conferences in Montreal in 1990 and Houston in 1991.[49] Would-be Mars explorers saw in this an opportunity . At the Case for Mars IV conference in June 1990, for example, Benton Clark suggested using the Energia heavy-lift rocket to transport Mars spacecraft propellants to orbit. "Use of the Soviet booster would, " he declared, "make the dependency between the cooperating countries simple and straightforward."[50] This represented a dramatic shift from the early 1980s, when Harrison Schmitt pushed for the LANL/NASA Manned Mars Missions study to help counter Soviet Mars moves.

In July 1990, Semyonov and Leonid Gorshkov, head of Energia's orbital stations department, published an article on Energia's Mars plans in the Soviet popular- audience publication Science in the USSR.[51] The con- figuration of the Mars spacecraft depended, they wrote, on the choice of "powerplant." They rejected chemical propulsion, saying that an all-chemical Mars ship would weigh upwards of 2,000 metric tons at Earth-orbit departure. A nuclear-thermal rocket Mars ship would weigh about 800 metric tons . More prom- ising, however, were solar-electric or nuclear-electric propulsion systems which could reduce ship mass to between 350 and 400 metric tons .

Semyonov and Gorshkov wrote that Soviet "aerospace technology is advanced enough to make a mission to Mars a reality," then summarized existing Soviet capa- bilities. In addition to the Energia rocket ("capable of lofting into Earth orbit whole sections of a spacecraft for final assembly"), the Soviet Union had "perfected the automatic docking procedures for putting together a spacecraft from sections in orbit" through more than 50 flights of automated Progress freighters to space stations. Semyonov and Gorshkov claimed that "[m]ost of the problems that would be faced by a crew on a long voyage to Mars in zero-gravity have been resolved" through 20 years of Soviet space station flights "in an environment very similar, if not identical, to that of a Mars mission." Finally, they reported that "electric . . . engines of the required parameters have been flaw- lessly performing on Earth."[52]

In 1991, Energia released a Mars expedition report reflecting "the expediency to take into account . . . world public opinion, which [is] against the launch of nuclear power"—an aversion reinforced by the Soviet Union's own April 1986 Chernobyl nuclear reactor meltdown.[53] NPO Energia's 355-metric-ton solar - electric Mars spacecraft would reach Earth orbit in sections strapped to the sides of five Energia heavy-lift rockets. The designers envisioned a pair of 40,000-square-meter solar panels supplying 7.6 megawatts of electricity at Earth's distance from the Sun and 3.5 megawatts at Mars.

The crew section of Energia's Mars ship design included two cylindrical modules linked end to end. The large living module would contain a "vitamin greenhouse" and individual cabins for four cosmonauts. Water tanks would surround the cabins to shield them from radiation. Over the course of the expedition the water would be gradually consumed and replaced by "waste bricks." An airlock for spacewalks and electric motors for pointing the solar arrays would separate the living module from a smaller control/laboratory module. The spacecraft's lithium-propellant electric propulsion system would be housed in twin modules attached to the sides of the control/lab module.

According to the report, Soviet designers had studied conical piloted Mars landers outwardly similar to the NAR MEM from 1969 to 1971.[54] Their 1991 Mars Landing Vehicle was, however, a cylinder with a conical forward section, a shape selected in part because it fit within the Energia rocket's payload envelope. The two-person lander's cylindrical section would house an ascent stage with a docking unit on top. The 60-metric-ton Mars Landing Vehicle would land horizontally. The cosmonauts would live in the lander's forward cone while on the Martian surface. After a week on Mars, the cosmonauts would blast off in the ascent stage to rejoin their comrades aboard the orbiting Mars ship.

At journey's end, the crew would separate from the Mars ship in the 10-metric-ton Earth Return Vehicle, a conical reentry module resembling the Apollo CM. The Earth Return Vehicle was designed for land landing—like the Soyuz space station transport, it would include solid-fueled soft-landing rockets under its ablative heat shield.

Bush and Gorbachev formally agreed at their July 1991 summit meeting to fly an American astronaut to Mir and a Soviet cosmonaut on the Space Shuttle. Less than two weeks later, in August 1991, communist hardliners launched an abortive coup d'etat against

Gorbachev, triggering the collapse of the Soviet Union. The following summer, Bush confirmed the July 1991 cooperation agreements with Russian President Boris Yeltsin. The first Russian cosmonauts arrived in Houston for Space Shuttle flight training in November 1992.

Space cooperation expanded dramatically under President William Clinton beginning in 1993. Space Station Freedom was redesigned as the International Space Station, which incorporated Russian hardware originally built for the Soviet Mir-2 space station. Mars-related cooperation, however, remained small in scale. For example, NASA Lewis researchers worked with Russian engineers on electric thrusters.

## America at the Threshold

The SEI Synthesis Group released its report America at the Threshold in May 1991.[55] The report, though written at a time when U.S.-Soviet space cooperation was becoming increasingly important to NASA's future, contained little on cooperation. The Synthesis Group report was the last in the series of high-profile documents proposing future directions for NASA that had begun with the National Commission on Space report in 1986.

Stafford headed a group of 22 experts from NASA and the Departments of Energy, Defense, and Transportation. They included retired JSC director Christopher Kraft and retired JSC engineering director Maxime Faget. Robert Seamans, retired from top NASA, Air Force, and Department of Energy posts, was Stafford's co-chair. They set up shop with a staff of 40 in Crystal City, Virginia, just outside Washington, DC.

The SEI Outreach Program provided the Synthesis Group with about 500 inputs from the 44,000-member AIAA. The Aerospace Industries Association, meanwhile, organized corporate briefings. These included a presentation by Martin Marietta featuring the Mars Direct plan. NASA took out newspaper advertisements around the country and set up toll-free telephone numbers to receive ideas from the public. About 900 concepts were submitted to Rand Corporation by early September. The national laboratories turned over their ideas during September. All told, the Synthesis Group had about 2,000 inputs in hand in late September.[56]

# Chapter 9: Space Exploration Initiative

The Synthesis Group was to submit at least two concepts based on these inputs to Truly, who would forward them to the National Space Council. A two-year NASA study would follow, during which the Agency would attempt to identify critical technologies needed to carry out the concepts proposed by the Synthesis Group.

In June 1991, the Group distributed 40,000 copies of its colorful report, emblazoned with the U.S. Presidential Seal, to industry, educators, government agencies, and international organizations. The report outlined four SEI architectures. In all of them, the ultimate goal was landing Americans on Mars. The Moon would serve as a rehearsal stage; nuclear systems would push spacecraft and power bases; and heavy-lift rockets would blast everything into orbit. Including nuclear propulsion was, as in the 1960s, in part a concession to Los Alamos, which had begun stumping for SEI nuclear systems as early as February 1990.[57]

In none of the architectures was Space Station Freedom an element of Mars transportation infrastructure. In September, Aviation Week & Space Technology quoted Stafford as saying, "I know when I went to the Moon . . . on Apollo 10, I did not have to stop at a space station."[58] This was a radical departure from SEI's ground rules. It was, in fact, a deviation from ground rules that had guided Mars planning since the time of the Apollo Moon missions, when NASA had first began to push for a space station.

Stafford's Architecture I emphasized Mars exploration but would spend five years on the Moon first. In 2005, a heavy-lift rocket would launch an automated cargo lander/habitat to the Moon. A second heavy-lift rocket would launch a crew of six to lunar orbit. Five astronauts would land on the Moon near the cargo lander; the sixth astronaut would mind the mothership in lunar orbit, just as the CM Pilot had minded his craft during Apollo Moon landing missions. The surface crew would stay on the Moon for 14 Earth days (one lunar daylight period).

In 2009-10, after four more heavy-lift rocket launches and two more lunar expeditions, a six-person Mars rehearsal crew would carry out a 300-day Mars expedition simulation in lunar orbit and on the Moon. After that, the Moon would not be visited again.

In 2012, the ninth heavy-lift rocket of Synthesis Group Architecture I would launch the first nuclear rocket of the program. It would push an automated cargo lander to Mars. The cargo lander would include a habitat identical to that landed on the Moon. The first six-person Mars crew would leave Earth in 2014 on the tenth heavy-lift rocket. After a flight lasting approximately 120 days, they would decelerate into Mars orbit using their nuclear-thermal rocket, separate from the Mars transfer habitat, and land near the 2012 cargo lander. The crew would spend 30 days testing systems and exploring before returning to the transfer spacecraft and firing the nuclear rocket for return to Earth. In the same launch opportunity, the eleventh heavy-lift rocket of the program would launch a cargo lander for the 2016 Mars expedition, which would spend 600 days on Mars. The report stated that Architecture I was conducive to more rapid execution (first Mars landing in 2008) if provided with "robust" funding.

The other architectures were generally similar. Architecture II, "science emphasis for the Moon and Mars," was designed to characterize the Moon and Mars scientifically through wide-ranging exploration and visits to multiple scientifically interesting landing sites. Architecture III, "Moon to stay and Mars exploration," emphasized a permanent lunar base. The base would achieve 18-person permanent staffing in 2007. A total of 47 six-person piloted expeditions would reach the Moon between 2004 and 2020, and the first piloted Mars landing would occur as in Architecture I.

The Stafford Group noted that "space is a unique store of resources: solar energy in unlimited amounts, materials in vast quantities from the Moon and Mars, gases from the [M]artian atmosphere, and the vacuum and zero gravity of space itself"—hence Architecture IV, which emphasized "space resource utilization."[59] Lunar ISRU would aim first for self-sufficiency; then it would export to Earth electricity and Helium-3 for fusion reactors. Mars ISRU would aim solely to provide self-sufficiency—the planet's greater distance would make exports to Earth impractical, the report stated. The Mars rehearsal on the Moon would take place as described in Architecture I, and Mars expeditions would occur in 2016 and 2018. The second expedition would establish an experimental greenhouse. Both expeditions would manufacture propellants for their rovers from Martian air.

The report made organizational recommendations for carrying out its program. It called upon NASA to establish "a long range strategic plan for the [N]ation's civil space pr ogram with the Space Explora tion Initiative as its centerpiece ," and asked President Bush to "establish a National Program Office by Executive order." In addition, it advocated advanced technology development programs.[60]

The SEI Synthesis Group had produced a cut-price version of The 90-Day Study—a disappointing outcome, given the magnitude of the Outreac h Program. Few Americans took notice of America at the Threshold, and

few of its recommendations were implemented. SEI funding fared no better in FY 1992 and FY 1993 than in the previous two years. The planned two-year follow-up study of critical technologies did not take place.

NASA disbanded the Headquarters Exploration Office in late 1992. The JSC Exploration Directorate c losed down a few months later.[61] The poorly attended Case for Mars V conference in May 1993 became SEI's wake. By the beginning of 1994, Mars planning across NASA threatened to slip back into its post-Apollo slumber.

# Chapter 10: Design Reference Mission

Recent developments in the exploration of Mars have served to focus attention once again on the possibilities for human expl o-ration of that planet. The unprecedented interest shown in the recently published evidence pointing to past life on Mars and in the Mars Pathfinder mission indicates that exploration of our solar system has not become so commonplace that the public cannot become surpris ed and fascinated by the discoveries being made . And these events have also rekindled the questions not of whether, but when will humans join the robots in exploring Mars . (Kent Joosten, Ryan Schaefer, and Stephen Hoffman, 1997)[1]

## Mars Direct

Like the STG, NCOS, and The 90-Day Study teams before it, the SEI Synthesis Group opted for a "brute-force" approach to piloted Mars exploration requiring such big-ticket items as heavy-lift rockets that dwarfed the old Saturn V, nuclear-thermal propulsion, and a lunar outpost. As has been seen, this approach has never gained muc h support. Proposing it repeatedly over the past 30 years has succeeded mainly in ingraining the belief that Mars exploration must be exorbitantly expensive (more expensive than a small w ar, for example) and needs decades to ac hieve its goal. Subsequent NASA Mars plans have sought to apply technologies new and old to reduce cost and tighten the schedule. They have begun the slow process of expunging the perception that a Mars mission must be conducted in a costly way.

Since 1992, NASA has based most of its Mars plans on the Mars Direct concept developed in 1990 by Martin Marietta. Mars Direct originated in Martin Marietta-sponsored efforts to develop SEI concepts. The plan has had staying power in part because it is an appealingly clever synthesis of concepts with respectable pedigrees. Mars Direct employs ISRU, aerobraking, a split mission architecture, a tether for artificial gravity, and a conjunction-class mission plan—all concepts that date from the 1960s or earlier . Mars Direct was influenced by the Case for Mars conferences , the Ride Report, and the NASA Exploration Office Studies , as well as ISRU researc h conducted by Robert Ash, Benton Clark, and others.[2]

Mars Direct has also had sta ying power since 1990 because one of its authors, engineer Robert Zubrin, has remained its zealous c hampion. On April 20, 1990, Zubrin and co-author Da vid Baker unveiled their plan to NASA engineers gathered at NASA Marshall.[3] Mars Direct went public at a National Space Society conference in Anaheim, California, in June 1990. It first received widespread attention a week later , after Zubrin presented it at the Case for Mars IV conference in Boulder, Colorado.[4]

In August 1990 the AIAA magazine Aerospace America carried a non-technical description of Mars Direct capturing Zubrin's promotional style.[5] It asked,

> Can the United States send humans to Mars during the present decade? Absolutely. We have developed vehicle designs and a mission architecture that can make this possible . Moreover, the plan we propose is not merely a "flags and footprints" one-shot expedition, but would put into place immediately an economical method of Earth-to-Mars transportation, vehicles for long-range surface exploration, and functional bases that could evolve into a mostly self-sufficient Mars settlement.[6]

Zubrin and Baker had the first Mars Direct expedition beginning in December 1996 with the launc h of a Shuttle-derived heavy-lift rocket from the Kennedy Space Center. The rocket, which Zubrin and Baker dubbed Ares, would consist of a modified Shuttle External Tank, two Advanced Solid Rocket Boosters, and four Space Shuttle Main Engines mounted on the External Tank's underside. A liquid hydrogen/liquid oxygen upper stage and an unpiloted Mars cargo lander covered by a streamlined shroud sat on top of the External Tank. The 40-ton cargo lander inc luded an aerobraking heat shield, descent stage, Earth-Return Vehicle, In-Situ Resource Utilization propellant factory, 5.8 tons of liquid hydrogen feedstoc k for propellant manufacture, and a 100-kilowatt nuclear reactor on a robot truck. The lander was, they wrote, "light enough for the booster upper stage to project it directly onto a six-month transfer orbit to Mars without any refueling or assembly in Earth orbit"—hence the name Mars Direct.[7]

The cargo lander would aerobrake in Mars' atmosphere and land. After touchdown, the robot truc k bearing the reactor would trundle a way to a natural depression or one created using explosives. It would lower the reactor into the crater—the crater rim would shield the landing site from radiation—then would run cables bac k to the lander. The reactor would activate, powering compressors which would draw in Martian air to manufacture propellant. Manufacturing propellants on Mars would help minimize the weight of propellants that had to be shipped from Earth.

The propellant factory would use the Sabatier process first proposed for use on Mars in 1978 by Robert Ash, William Dowler, and Giulio Varsi. Liquid hydrogen feedstock would be exposed to Martian carbon dioxide in the presence of a catalyst, producing 37.7 tons of methane and water. The methane would be stored and the water electrolyzed to yield oxygen and more hydrogen. The oxygen would then be stored and the hydrogen recycled to manufacture more water and methane. Additional oxygen would be manufactured by decomposing carbon dioxide into carbon monoxide and oxygen and venting the carbon monoxide. In a year, the propellant factory would manufacture 107 tons of methane and oxygen propellants. The piloted Mars spacecraft would not be launched until the automated cargo ship finished manufacturing the required propellants, thereby reducing risk to crew.

In January 1999—the next minimum-energy Mars transfer opportunity—two more Ares rockets would lift off. One would carry a cargo lander identical to the one already on Mars. The other would carry a "manned spacecraft looking somewh at like a giant hockey puck 27.5 f[ee]t in diameter and 16 f[ee]t tall" based on Martin Marietta desig ns developed for the NASA Office of Exploration. [8] The top floor would comprise living qua rters for the four -person crew, while the bottom floor would be stuffed with cargo and equipment, including a pressurized rover. Zubrin and Baker estimated the piloted spacecraft's weight at 38 tons.

The upper stage would launch the "hockey puck" spacecraft on course for Mars and separate, but the two would remain attached by a 1,500-meter tether. This assemblage would rotate once per minute to produce acceleration equal to Martian surface gravity in the piloted spacecraft. A similar lightweight artificial gravity concept w as proposed by Robert Sohn in

1964. Near Mars the upper stage and tether would be discarded.

The piloted spacecraft would aerobrake into Mars orbit, then land near the 1996 cargo lander. No part of the ship would remain in orbit. Landing the entire crew on the surface would help minimize risk. Once on Mars, the Martian atmosphere would provide some radiation protection, and the crew could use Martian dirt as additional shielding. They would also experience Martian gravity. Though only a third as strong as Earth's gravity, it seemed likely that even that small amount would be preferable to a long weightless stay in Mars orbit.

As in the SAIC split-sprint plan, the crew would have to rendezvous at Mars with propellants for their trip home. This was seen by some as increasing risk. Unlike the SAIC crew, however, the Mars Direct astronauts would have options if they could not reach their Earth-return propellants.

Baker and Zubrin pointed out that the crew had their rover to drive to the 1996 cargo lander, though ideally they would land within walking distance. If some gross error meant they landed more than 600 miles from the 1996 cargo lander—beyond the range of their rover—they could command the cargo lander launc hed with them in 1999 to land nearby. It would then manufacture propellant for their return to Earth. If the 1999 cargo lander failed, the Mars Direct astronauts would have sufficient supplies to hold out until a relief expedition arrived in two years. Assuming that the crew landed near the 1996 cargo lander as planned, the 1999 cargo lander would set down 500 miles from the first Mars landing site and begin to make propellants for the second Mars expedition, which would leave Earth in 2001.

Eleven of the 107 tons of propellants manufactured by the 1996 cargo lander would be set aside to power the pressurized rover. During their 500-da y stay on Mars, the explorers would conduct long tra verses—up to 600 miles round-trip—thoroughly characterizing the region around their landing site. This impressive capability would maximize science return by allowing the crew to survey large areas, though with some increased risk. If the rover broke down, the crew could become stranded beyond hope of rescue, hundreds of kilometers from base.

At the end of the 500-da y Mars sta y, the ERV engine would ignite, burning methane and oxygen propellants manufactured using the Martian atmosphere . The small ERV spacecraft would use the cargo lander as a launch pad to perform ascent and direct insertion onto a trajectory to Earth. After six weightless months in the cramped ERV, the crew would reenter Earth' s atmosphere and perform a parac hute landing. The small ERV was considered by many to be a weak link in the Mars Direct plan.

The 2001 expedition crew would land near the 1999 cargo lander. If all went as planned, the 2001 cargo lander would land 500 miles a way. The 2003 crew would land next to the 2001 cargo lander , while the 2003 cargo lander would touch down 500 miles away for the 2005 expedition, and so on. After several expeditions, a network of bases would be established. "Just as towns in the western U.S. grew up around forts and outposts," wrote Baker and Zubrin, "future [M]artian towns would spread out from some of these bases . As information returns about eac h site, future missions might return to the more hospitable ones and larger bases would begin to form."[9]

## SEI's Last Gasp

In SEI's last days, the Stafford Synthesis Group report formed the basis of NASA's Mars planning. From 1991 to 1993, the Agency performed the First Lunar Outpost (FLO) study, which took as a point of departure the lunar elements of the Synthesis Group' s four architectures. In the summer of 1992, the NASA Headquarters Exploration Office under Michael Griffin, the successor to the Office of Exploration first headed by Sally Ride , launched a NASA-wide study to determine how FLO might find hardw are commonality with a follow-on Mars expedition, thereby reducing the costs of both programs.[10]

The Mars Exploration Study Team workshop held in August 1992 produced a plan containing elements of both Mars Direct and the Synthesis Group Mars plan. It was briefed to Griffin in September.[11] The May 1993 Mars Exploration Study Team workshop produced a Mars expedition Design Reference Mission (DRM) with little overt FLO commonality beyond a common heavy-lift rocket and outwardly similar vehicles for lunar and Mars ascent. In fact, the DRM was modeled

on Mars Direct. Robert Zubrin was an advisor to the Mars Exploration Study Team in late 1992 and 1993. He briefed Griffin on Mars Direct in June 1992, then briefed the JSC Exploration Program Office in October 1992.[12]

The Mars Exploration Study Team DRM was reported in a workshop summary and in tec hnical papers in September and November 1993.[13, 14] It included the following:

- no low-Earth orbit operations or assembly—that is, no reliance on a space station as a Mars transportation element,

- no reliance on a lunar outpost or other lunar operations,

- heavy-lift rocket capable of launching 240 tons to low-Earth orbit, 100 tons to Mars orbit, and 60 tons to the Martian surface (more than twice the capability of the Saturn V),

- short transit times to and from Mars and long Mars surface sta y times beginning with the first expedition (conjunction-class missions),

- six crewmembers to ensure adequate manpower and skills mix,

- early reliance on Mars ISRU to minimize weight launched to Mars, and

- common design for surface and transit habitats to reduce development cost.

The most significant difference between Mars Direct and the Mars Exploration Study Team's DRM was the division of the Mars Direct ERV functions between two vehicles. In the Mars Direct plan, the ERV lifted off from Mars at the end of the surface mission and flew directly to Earth. In the judgment of many , however, the Mars Direct ERV was too small to house four astronauts during a six-month return from Mars , let alone the DRM's six astronauts.[15] In the DRM, therefore, only a small Mars Ascent Vehicle (MAV) would rely on ISRU. The crew would use it to reach Mars orbit at the end of their surface sta y and doc k with the orbiting ERV. The addition of a rendezvous and docking in Mars orbit was seen by some as increasing risk to crew , but

# Chapter 10: Design Reference Mission

Figure 23—NASA's 1993 Mars mission plan: after landing on Mars, the automated propellant factory manufactures liquid methane and liquid oxygen propellants f or the conical Mar s Ascent Vehicle it carries on top. (NASA Photo S93-50643)

Figure 24—The crew Habitat lands near the propellant factory with empty propellant tanks. Note wheels for moving the Habitat on the martian surface. (NASA Photo S93-050645)

there seemed to be little alternative if a realistically large ERV was to be provided.

The September 2007 Mars transfer opportunity w as used for the study because it would be c hallenging in terms of time and energy required for Mars transfer , not necessarily because an expedition w as planned for that time. The first expedition would begin with launch of three heavy-lift rockets, each bearing one unmanned spacecraft and one nuclear propulsion upper stage. The three spacecraft were the cargo lander, the ERV orbiter, and an unmanned Habitat lander . They would weigh between 60 and 75 tons each, a weight estimate considered more realistic than the 30 to 40 tons quoted in Mars Direct.

The ERV and Habitat designs were based on a common crew module design resembling the Mars Direct "hockey puck." The cargo lander would carry the MAV, ISRU propellant factory, and hydrogen feedstock, along with 40 tons of cargo , including the pressurized rover . All would reach Mars during August and September 2008. The ERV would aerobrake into Mars orbit, while the cargo lander and Habitat would land on Mars . The cargo lander would then set about manufacturing 5.7 tons of methane and 20.8 tons of oxygen for the MA V and a 600-day cache of life-support consumables.

As in Mars Direct, the crew would follow during the next Mars launch opportunity 26 months later (October - November 2009), accompanied by unmanned vehic les supporting the next expedition or providing bac kup for those already on Mars. The explorers would land near the 2007 cargo lander and Habitat. The Habitats would

Figure 25—Mars Base 1: the crew docks its Habitat on the surface with a second Habitat and begins a 600-day stay. They use a pressurized rover (left) to explore up to 500 kilo - meters from base. (NASA Photo S93-45582)

include wheels to allow the explorers to move them together so they could be linked using a pressurized tunnel. The 2007 Habitat would also provide a bac kup pressurized volume if the 2009 Habitat w as damaged during landing and rendered uninhabitable.

The first Mars outpost thus established, the crew would unpack the pressurized rover from the 2007 cargo lander. During their 600-da y stay on Mars, the crew would carry out several 10-da y rover tra verses ranging up to 500 kilometers from the outpost.

In October 2011, the 2009 crew would lift off from Mars in the 2007 MAV. They would dock in Mars orbit with the 2007 ERV and fire its twin liquid methane/liquid oxygen rocket engines to lea ve Mars orbit for Earth, retaining the MAV capsule. Near Earth the explorers

Figure 27—Mars Orbit Rendezvous: The Mars Ascent Vehicle docks with the Earth Return Vehicle in Mars orbit. The Earth Return Vehicle's rocket engines would place the crew on a six-month low-energy trajectory homeward. (NASA Photo S93-27626)

Figure 26—Using the propellant factory as a launc h pad, the Mars Ascent Vehicle blasts off burning propellants made from terrestrial hydrogen and Martian atmospheric carbon dioxide. (NASA Photo S93-050644)

would enter the MA V capsule and detac h from the ERV, which would sail past Earth into solar orbit. They would then reenter Earth's atmosphere and perform a parachute landing.

The Mars Exploration Study Team effort was SEI's last gasp. Before it was completed, NASA had begun to dismantle its formal Mars exploration planning organization. The Headquarters Exploration Office w as abol-

ished in late 1992. The JSC Exploration Directorate , created soon after T he 90-Day Study' s release, was trimmed back and re-created as the JSC Planetary Projects Office.[16]

As the apparatus for piloted Mars planning within NASA shrank, automated Mars exploration also suffered a cruel blow . Mars Observer, the first U.S. automated Mars mission since the Vikings, had left Earth on 25 September 1992. On 21 August 1993, three days before planned Mars orbit arrival, the spacecraft's transmitter was switched off as planned to protect it from shocks during propellant system pressurization. Contact was never restored. An independent investigation report released in January 1994 pointed to a propulsion system rupture as the most probable cause of Mars Observer' s loss, the first post-launc h failure of a U .S. planetary exploration mission since Surveyor 4 in 1967. [17]

NASA almost immediately announced plans to fly Mars Observer's science instruments on an inexpensive Mars orbiter as soon as possible . This marked the genesis of the Mars Surveyor Program, which aimed to launc h low-cost automated spacecraft to Mars every 26 months at each minimum-energy launch opportunity.[18]

## Refreshed Dreams

In 1994, the JSC Planetary Projects Office , NASA's de facto focus for piloted Mars planning following abolition of the Headquarters Exploration Office, was downsized, then abolished. In February it became a branch of the JSC Solar System Exploration Division, and in June its remaining personnel were assigned to the JSC Office of the Curator, where they explored low-cost options for sending people to the Moon.[19] The Curator's Office managed disposition of Apollo lunar samples and meteorites, including one meteorite designated ALH 84001. Even as the Planetary Projects Office was abolished, ALH 84001 was determined to have originated on Mars.

On 7 August 1996, NASA, Stanford University, and McGill University scientists led by NASA scientist David McKay announced that they had discovered possible fossil microorganisms in Martian meteorite ALH 84001. In a NASA Headquarters press conference , the McKay team cited the evidence for past Martian life. This included the presence of complex carbon compounds resembling those produced when Earth bacteria die, magnetite particles similar to those in some Earth bacteria, and segmented features on the scale of some Earth nanobacteria. McKay told journalists,

> There is not any one finding that leads us to believe that this is evidence of past life on Mars Rather, it is a combination of many things that we have found. They include Stanford's detection of an apparently unique pattern of organic molecules, carbon compounds that are the basis of life. We also found several unusual mineral phases that are known products of primitive microorganisms on Earth. Structures that could be microscopic fossils seem to support all of this The relationship of these things in terms of location—within a few hundred-thousandths of an inch of eac h other—is the most compelling evidence.[20]

According to their analysis, the 1.9-kilogram rock soaked in carbonate-rich water containing the possible microorganisms 3.6 billion years ago . It lay in the Martian crust, shocked by the occasional local upheaval, until an asteroid impact blasted it off Mars 16 million years ago. After orbiting the Sun several million times, ALH 84001 landed in Antarctica 13,000 years ago, where it was collected on 27 December 1984 in the Allan Hills ice field.[21]

The McKay team's discovery generated unprecedented public enthusiasm for Mars, which in turn provided the catalyst for reestablishment of the JSC Exploration Office in November 1996. The new office, managed by Doug Cooke, was reconstituted as part of the Advanced Development Office in the JSC Engineering Directorate.[22] Mars planners dusted off the 1993 DRM to serve as the point of departure for new planning .

At the same time , NASA Headquarters took an important step toward eventual piloted Mars exploration. On 7 November 1996, Associate Administrator for Space Flight Wilbur Trafton, Associate Administrator for Space Science Wesley Huntress, and Associate Administrator for Life and Microgra vity Sciences and Applications Arnauld Nicogossian signed a joint memorandum calling for NASA's Human Exploration and Development of Space (HEDS) Enterprise and Space Science Enterprise to work together toward landing humans on Mars.

They told J et Propulsion Laboratory director Edw ard Stone and JSC director George Abbey that "[r]ecent developments regarding Mars and the growing maturity of related programs lead us to believe that this is the right time to fully integrate several areas of robotic and human Mars exploration study and planning ."[23] The Associate Administrators then gave Stone and Abbey until 1 F ebruary 1997, to produce "a proposal that NASA can bring forw ard, after successful deployment of the International Space Station, for human exploration missions beginning sometime in the second decade of the next [21st] century."[24]

Trafton, Huntress, and Nicogossian also asked for "a credible approach to achieving affordable human Mars exploration missions." They defined "a credible cost" as "the amount currently spent by NASA on the International Space Station"—that is, less than $2 billion annually. This was a dramatic reduction over the $15 billion per year proposed in the excised cost section of The 90-Day Study. They asked that Stone and Abbey identify "technology investments and developments that could dramatically decrease the cost of human and robotic missions."[25]

In March 1997, the HEDS and Space Science Enterprises agreed that the 2001 Mars Surveyor lander should include instruments and technology experiments supporting piloted Mars exploration. Among the planned experiments was a compact system for testing ISRU propellant manufacture on Mars. In a press conference, Huntress called it "the first time since the 1960s" that "NASA's space science and human space flight programs are cooperating directly on the exploration of another planetary body." Trafton called the joint effort "a sign that NASA is acquiring the information that will be needed for a national decision, perhaps in a decade or so, on whether or not to send humans to Mars."[26]

In addition to stating that NASA 's robotic program would complement its piloted Mars flight planning efforts, the joint memorandum showed that, at a high managerial level, NASA had not abandoned its plans to eventually send people to Mars despite SEI' s collapse. There was no firm timetable for accomplishing the piloted Mars mission and no Presidential dec laration. Instead, there was a new philosophy—continuing low-level, low-cost planning, much of it in-house , and low-level Earth-based technology research accompanied by efforts to use the existing low-cost robotic exploration program to answer questions relevant to piloted exploration. In short, the Agency accepted public ly for the first time that it might eventually send people to Mars without recourse to a new large program—without a new Space Exploration Initiative or Apollo program. This philosophy continues to guide NASA Mars planning at the time of this writing (mid-2000).

Success or failure in the automated Mars program thus became success or failure for piloted Mars planners . The joint human-robotic Mars effort received a boost on 4 July 1997, when Mars Pathfinder successfully landed at Ares Vallis, one of the large outw ash channels first spotted by Mariner 9 in 1971 and 1972. Pathfinder, the first U.S. Mars lander since the Vikings, dropped to the rock-strewn surface and bounced to a stop on airbags , then opened petals to right itself and expose instruments and solar cells. The technique was similar to the one the Soviets employed to land robots on the Moon in the 1960s and on Mars in the 1970s. The Sojourner rover—the first

automated rover to operate on another world since the Soviet Union's Lunokhod 2 explored the Moon in 1972—crawled off its perc h on one of P athfinder's petals and crept about the landing area analyzing rock and dirt composition. Sojourner and Pathfinder—the latter renamed the Sagan Memorial Station—successfully completed their primary mission on 3 August.

As Mars Pathfinder bounced to a successful landing in Ares Vallis, the glossy report Human Exploration of Mars: The Reference Mission of the NASA Mars Exploration Study Team rolled off the presses.[27] In addition to a detailed description of the 1993 DRM, the July 1997 document cont ained general recommendations on the conduct of a piloted Mars program based on experience gained through SEI and the Space Station program.

The report recommended that NASA set up "a Mars Program Office . . . early in the process ." It also proposed to avoid Space Station's redesigns and delays by establishing "a formal philosophical and budgetary agreement . . . as to the objectives and requirements imposed on the mission before development is initiated, and to agree to fund the project through to completion." Finally, taking into account the McKay team's discovery, it called for "adequate and acceptable human quarantine and sample handling protocols early in the Mars exploration program" to protect Earth and Mars from possible biological contamination.[28]

The JSC Exploration Office called its report "another chapter in the ongoing process of melding new and existing technologies, practical operations, fiscal reality, and common sense into a feasible and viable human mission to Mars," adding that "this is not the last chapter in the process, but [it] marks a snapshot that will be added to and improved upon by others in the future."[29] In fact, by the time the report sa w print, the next chapter was nearly complete.

## Scrubbing the DRM

Subsequent DRM evolution focused on minimizing spacecraft weight in an effort to reduce estimated mission cost. The slang term engineers used to describe this process w as "scrubbing." The 1997 "scrubbed" DRM went public in August 1997.[30] It minimized mass by reducing common Habitat diameter; combining the functions of the pressure hull, aero-

Figure 28—NASA's 1997 Mars plan proposed to reduce weight by using an aerobrake integrated with the spacecraft hull and nuclear roc kets. These steps would help eliminate need for a heavy-lift roc ket, permitting a cheaper Shuttle-derived launch system. (NASA Photo S97-07844)

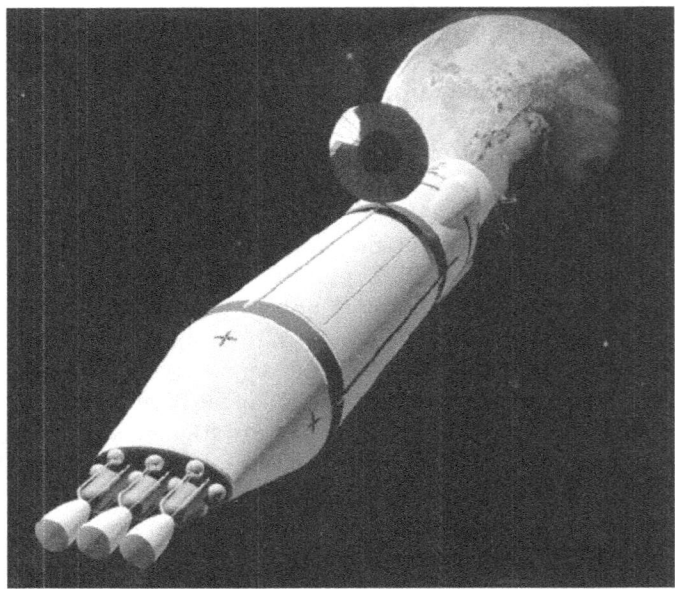

Figure 29—Nuclear stages in NASA 's 1997 Mars plan included engines (left) based on revived 1960s NER VA technology. (NASA Photo S97-07843)

brake heat shield, and Earth launc h shroud; and employing lightweight composite structures . The nuclear stages for injecting the spacecraft tow ard Mars would be launc hed into Earth orbit without spacecraft attached, then docked with the spacecraft in Earth orbit. These steps and others allowed planners to eliminate the 1993 DRM's large heavy-lift rocket, potentially the costliest mission element.

To place the first crew on Mars , the 1997 DRM would require eight launc hes of a Shuttle-derived roc ket capable of boosting 85 tons into Earth orbit. In the first launch opportunity, six of these roc kets would launc h payloads—three nuclear propulsion stages and three Mars spacecraft (cargo lander , ERV, and unpiloted Habitat). Each spacecraft would doc k with its nuc lear stage in Earth orbit, then launch toward Mars. In the second launch opportunity, 26 months later, six more Shuttle-derived rockets would launc h three nuc lear stages and three spacecraft, including a Habitat lander containing the crew. The spacecraft would dock with their nuclear stages and launc h toward Mars. The rest of the mission plan closely resembled the 1993 DRM. To accomplish the first expedition, the 1997 DRM would launc h 303 tons to Mars—75 tons less than the 1993 DRM.

The new DRM was on the street, and a few weeks later, a new automated spacecraft w as orbiting Mars. On 11 September 1997, the Mars Global Surveyor orbiter, the first spacecraft in the Mars Surveyor Program, arrived in an elliptical Mars orbit after a 10-month flight. Mars Global Surveyor carried bac kups of instruments lost with Mars Observer in 1993. It commenced a series of passes through Mars' upper atmosphere to reac h a lower, more circular Mars orbit without using propellants. A damaged solar arra y threatened to collapse under the pressure of atmospheric drag , however, so the aerocapture maneuvers had to be extended over a year. Nevertheless, the spacecraft turned its instruments toward Mars and began initial observations .

## Defining the Surface Mission

As Mars planners sought to minimize spacecraft weight, it became clear that they would require more data on the mission' s Mars surface pa yload. Planners historically have spent little time detailing what astronauts would do once they landed on Mars. To begin the process of better defining the 500-to-600-day Mars surface mission, veteran Moon and Mars planner Mic hael

Duke chaired a workshop held at the Lunar and Planetary Institute in Houston on 4-5 October 1997. [31]

Workshop participants divided into two workin g groups. The Science and Resources group based its discussions on a "three-pronged approach" to Mars exploration. Mars explorers would seek evidence of life or its precursors and attempt to understand Mars c limate history. They would also act as prospectors , seeking water, minerals, energy, and other resources for supporting future Mars settlements . This three-pronged science approach also guided the automated Mars Surveyor program. [32]

The Living and Working on Mars group looked at chores the crew would need to perform during their Mars stay. These included initial base setup, such as deploying an inflatable greenhouse, and base maintenance, such as ridding air filters of ever-present ultra-fine Martian dust. Astronauts on Mars would also harvest crops , service their space suits , and perform less mundane tasks suc h as exploring the surface in the pressurized rover and drilling deep in search of Martian microorganisms that might hide far beneath the surface.

The workshop recommended that "a process and program be put into place whereby a wide range of people could contribute to the thought process ." The report urged that students in particular be involved, because "their representatives will be the ones who are actually to do this exploration." [33]

## A New Concept

Meanwhile, engineers at NASA Lewis studied using solar-electric propulsion in the DRM to further reduce the amount of weight that would have to be launched into orbit. In January 1999, they proposed a novel concept using a Solar -Electric Transfer Vehicle (SETV) which never left Earth orbit, but which provided most of the energy needed to launc h the Mars vehicles from Earth orbit toward Mars. [34]

The 1997 DRM required eight Shuttle-derived roc kets for the first Mars expedition. By contrast, the Lewis solar-electric DRM required only five roc kets. Removal of the bac kup Habitat lander—a decision taken by Mars planners in the JSC Exploration Program

Office—eliminated two hea vy-lift rockets. Replacing the four nuclear stages used to leave Earth in the 1993 and 1997 DRMs with the SETV and three small expendable chemical stages eliminated one more . This substitution also eliminated the cost of developing a nuclear rocket engine and the potential political headaches of launching nuclear payloads.

The Lewis team envisioned a self-erecting SETV weighing 123 tons and measuring 194.6 meters across its thin-film solar arra ys. The arrays would provide electricity to two sets of Stationary Plasma Thrusters (SPTs), also known as TAL (Thruster with Anode Layer) or Hall thrusters , an electric propulsion tec h-nology pioneered by the Russians.

The SETV would need months to complete large orbit changes. Because of this , it would spend considerable time crossing through Earth's Van Allen Radiation Belts. This meant that the Lewis DRM vehicles would require radiation-hardened systems . The authors assumed that the SETV would be good for two missions beyond the Van Allen belts before radiation, temperature extremes, meteoroid impacts, and ultraviolet light seriously degraded its solar arrays.

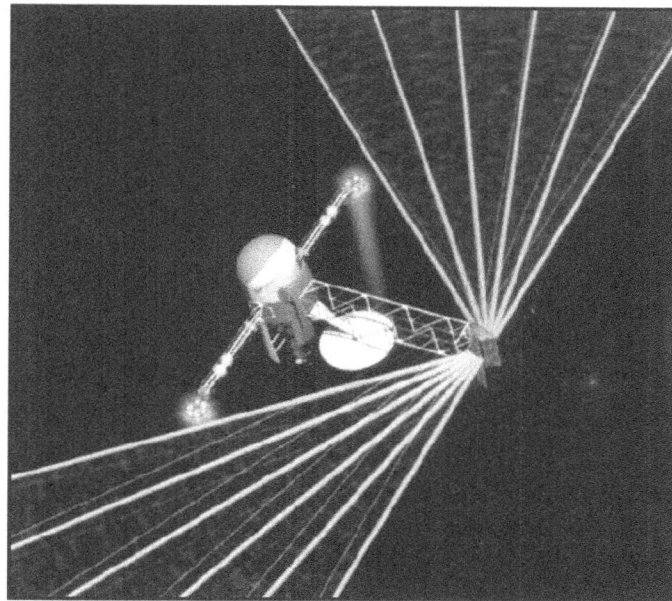

Figure 30—In 1998, NASA Lewis Research Center proposed a reusable Solar-Electric Transfer Vehicle (SETV) and clever use of orbital mec hanics to reduce Mars expedition mass. SETV's solar panel spars would inflate in orbit, spreading "wings" of solar cell fabric. (NASA Photo S99-03585)

The SETV's first mission would place one unpiloted cargo vehicle and one unpiloted ERV, each with a small chemical rocket stage, into High-Energy Elliptical Parking Orbit (HEEPO) around the Earth. The SETV would start in a nearly circular low-Earth orbit and raise its apogee by operating its SPT thrusters only at perigee. It would need from six to twelve months to raise its apogee to the proper HEEPO for Earth-Mars transfer. The final HEEPO apogee would be more than 40,000 kilometers, making it very lightly bound by Earth's gravity.

When Earth, Mars, and the plane of the HEEPO were properly aligned for Earth-Mars crossing, the SETV would release the cargo lander, ERV, and small chemical stages. At next perigee the chemical stages would ignite, pushing the spacecraft out of the HEEPO on a path that would intersect Mars six months later. After releasing the chemical stages and spacecraft, the SETV would point its SPTs in its direction of motion and operate them at perigee to return to a circular low-Earth orbit.

The SETV's second mission would place one Habitat lander with a small chemical stage into HEEPO. Because the climb to HEEPO again would require up to twelve months and long periods inside the Van Allen Radiation Belts, the Habitat lander would remain unpiloted until just before Earth orbit departure. As the SETV climbed toward planned final HEEPO apogee, a small, chemical-propellant "taxi" carrying the Mars crew would set out in pursuit. The crew would transfer to the Habitat lander, cast off the taxi, then separate the Habitat lander and chemical stage from the SETV. At the next perigee, the chemical stage would ignite to place the first expedition crew on course for Mars. The remainder of the first Mars expedition would occur as described in the 1997 scrubbed DRM, except for the absence of a backup Habitat lander.

In February 1999, soon after the Lewis team made public their variation on the 1997 DRM, Mars Global Surveyor achieved its nominal mapping orbit. At this writing, exploration and data interpretation are ongoing, but it is already clear that the spacecraft is revolutionizing our understanding of Mars. By mid-2000, its instruments had detected evidence that Mars once had a strong planetary magnetic field, a finding potentially important for the early development of Martian life; that Mars' polar regions once knew extensive glaciers; and that water flowed on Mars' surface recently, and perhaps flows occasionally today, carving gullies in cliffs and crater walls.

## Not the Last Chapter

In May 1998, a small team of NASA and contractor space suit engineers traveled to sites in northern Arizona where Apollo Moonwalkers had trained three decades before. They observed and assisted as a veteran geologist wearing a space suit performed geological field work and set out simulated scientific instruments in Mars-type settings—for example, on the rim of Meteor Crater. The team contained cost by traveling from Houston to Arizona overland and by reusing a space suit originally designed for Space Station Freedom. In addition to gathering data on space suit mobility to enable design of future Mars space suits, the exercise permitted veteran space suit engineers who had participated in the development of the Apollo lunar space suits to pass on their experience to young engineers who had been children, or not yet born, when Americans last walked on an alien world. [35]

Michael Duke and the other organizers of the Human Exploration and Development of Space-University Partners (HEDS-UP) program had a similar motive. They sought to involve and inspire the next generation of Mars planners, who might become the first generation of Mars explorers. In May 1998, the first HEDS-UP Annual Forum saw undergraduate and graduate design teams from seven universities across the United States present Mars design studies.[36] Twice as many universities sent enthusiastic students to the 1999 HEDS-UP Annual Forum.[37]

In the nearly half-century since von Braun wowed Americans with visions of Mars flight in Collier's magazine, our understanding of Mars has steadily improved. We have progressed from hazy telescopic views of Mars to pictures on the Internet of Sojourner rearing up on a flood-tossed Martian boulder. Plans for piloted Mars exploration have matured in step with our improved vision. For example, no longer do planners seek to bring all necessities from Earth, for now it is known that Mars has useful resources.

The Mars planning concepts developed in the twilight years of the second millennium form a launch pad for Mars planners—and perhaps Mars explorers—at the dawn of the third. Current technological trends—for example, increasingly capable miniaturized robots and direct public engagement in Mars exploration through the Internet—promise to reshape Mars planning.

Yet it should be remembered that ISRU, the concept that dominated Mars planning in the 1990s, dates from the 1960s and 1970s. This suggests that, in addition to whatever new revolutions future technological development brings, other revolutions might lie buried in the historical archives awaiting the careful and imaginative researcher. Further, this suggests that Mars planners should carefully preserve their work lest they deprive future planners of useful concepts.

Young people now looking to Mars, such as the student participants in the HEDS-UP program, should not have to waste their time reinventing old concepts. They should instead be able to study the old concepts and build new ones upon them. They should also be able to study the political and social settings of the old concepts, so that they might better navigate the "illogical" pitfalls that can bring down a technically logical Mars plan. Providing the next generation with the history of Mars planning helps hasten the day when humans will leave bootprints on the dusty red dunes of Mars.

# Acronyms

AAP           Apollo Applications Program
AAS           American Astronautical Society
ABMA        Army Ballistic Missile Agency
AEC           Atomic Energy Commission
AIAA         American Institute of Astronautics and Aeronautics
CIA           Central Intelligence Agency
CM           Command Module
CPS           Chemical Propulsion Stage
CSM         Command and Service Module
DRM        Design Reference Mission
EMPIRE    Early Manned Planetary-Interplanetary Roundtrip Expeditions
EOR          Earth-Orbit Rendezvous
ERV          Earth-Return Vehicle
ET            External Tank
FLEM         Flyby-Landing Excursion Mode
FLO          First Lunar Outpost
FY            Fiscal Year
GNP         Gross National Product
HEDS         Human Exploration and Development Space
HEDS-UP   Human Exploration and Development Space—University Partners
HEEPO     High-Energy Elliptical Parking Orbit
IPP           Integrated Program Plan
ISRU         In-Situ Resource Utilization
ISV          Interplanetary Shuttle Vehicle
JAG         Joint Action Group
JPL           Jet Propulsion Laboratory
JSC          Johnson Space Center
KSC          Kennedy Space Center
LANL         Los Alamos National Laboratory
LLNL         Lawrence Livermore National Laboratory
LOR         Lunar-Orbit Rendezvous
LSS          Life Support Section
M             Maneuver
MAV         Mars Ascent Vehicle
MEM        Mars Excursion Module
MEV         Mars Exploration Vehicle, Mars Excursion Vehicle
MMM       Manned Mars Mission
MOR        Mars-Orbit Rendezvous
MSC         Manned Spacecraft Center
MSSR        Mars Surface Sample Return
NACA        National Advisory Committee for Aeronautics
NAR         North American Rockwell
NASA         National Aeronautics and Space Administration
NCOS       National Commission on Space
NERVA     Nuclear Engine for Rocket Vehicle Application
NRC         National Research Council
NRDS        Nuclear Rocket Development Station
OMB        Office of Management and Budget
OMS        Orbiter Maneuvering System

# Acronyms

| | |
|---|---|
| OMSF | Office of Manned Space Flight |
| OTV | Orbital Transfer Vehicle |
| PH-D | Phobos-Deimos |
| PM | Propulsion Module |
| PMRG | Planetary Missions Requirements Group |
| PPM | Primary Propulsion Module |
| PSAC | President's Science Advisory Committee |
| REM | Roentgen Equivalent Man |
| RIFT | Reactor-In-Flight Test |
| SAIC | Science Applications International Corporation |
| SEI | Space Exploration Initiative |
| SETV | Solar-Electric Transfer Vehicle |
| SM | Service Module |
| SNPO | Space Nuclear Propulsion Office |
| SOC | Space Operations Center |
| SPT | Stationary Plasma Thrusters |
| SRB | Solid Rocket Booster |
| SSME | Space Shuttle Main Engine |
| STG | Space Task Group |
| STS | Space Transportation System |
| TAL | Thruster with Anode Layer |
| UMPIRE | Unfavorable Manned Planetary-Interplanetary Roundtrip Expeditions |
| WGER | Working Group on Extraterrestrial Resources |

# Endnotes

## Preface

1. Edward Ezell, "Man on Mars: The Mission That NASA Did Not Fly" (paper presented at the American Association for the Advancement of Science Annual Meeting, Houston, Texas, 3-8 January 1979), p. 24.

2. Readers seeking additional information on Mars planning are directed to the author's Web site Romance to Reality (http://members.aol.com/dsfportree/explore.htm), which contains over 250 annotations of Moon and Mars planning documents, with more added regularly.

## Chapter 1

1. Wernher von Braun with Cornelius Ryan, "Can We Get to Mars?" *Collier's* (30 April 1954), p. 23.

2. Frederick Ordway and Mitchell Sharpe, *The Rocket Team* (New York: Thomas Y. Crowell, 1979), p. 408.

3. Wernher von Braun, *The Mars Project* (Urbana, IL: University of Illinois Press, 1962).

4. *Ibid.*, p. 3.

5. *Ibid.*, p. 75.

6. Louise Crossley, *Explore Antarctica* (Cambridge, England: Cambridge University Press, 1995), p. 40.

7. Fred Whipple and Wernher von Braun, "Man on the Moon: The Exploration," *Collier's* (25 October 1952), p. 44.

8. Wernher von Braun, "Crossing the Last Frontier," *Collier's* (22 March 1952): 24-29, 72.

9. Wernher von Braun, "Man on the Moon: The Journey," *Collier's* (18 October 1952): 52-60; Whipple and von Braun, "Man on the Moon: The Exploration," pp. 38-48.

10. Von Braun with Ryan, "Can We Get to Mars?" pp. 22-28.

11. *Ibid.*, pp. 26-27.

12. Willy Ley and Wernher von Braun, *The Exploration of Mars* (New York: Viking Press, 1956).

13. *Ibid.*, p. 85.

14. *Ibid.*, p. 98.

15. *Ibid.*, p. 157.

## Chapter 2

1. John F. Kennedy, "Excerpts from 'Urgent National Needs,' " Speech to a Joint Session of Congress, 25 May 1961, in John Logsdon, gen. ed., with Linda Lear, Janelle Warren-Findlay, Ray Williamson, and Dwayne Day, *Exploring the Unknown: Selected Documents in the History of the U.S. Civil Space Program, Volume I: Organizing for Exploration* (Washington, DC: NASA SP-4407, 1995), pp. 453-54.

# Endnotes

2. Robert Merrifield, "A Historical Note on the Genesis of Manned Interplanetary Flight," AAS Preprint 69-501 (paper presented at the AAS 15th Annual Meeting, 17-20 June 1969), p. 7.

3. David S. F. Portree, *NASA's Origins and the Dawn of the Space Age* (Washington, DC: NASA Monographs in Aerospace History #10, 1998), pp. 8-11.

4. Ezell, "Man on Mars," pp. 5-6; see also Merrifield, "A Historical Note," p. 8.

5. S. C. Himmel, J. F. Dugan, R. W. Luidens, and R. J. Weber, "A Study of Manned Nuclear-Rocket Missions to Mars," IAS Paper No. 61-49 (paper presented at the 29th Annual Meeting of the Institute of Aerospace Sciences, 23-25 January 1961), p. 2.

6. *Ibid.*, p. 5.

7. *Ibid.*, p. 18.

8. Von Braun with Ryan, "Can We Get to Mars?" p. 24.

9. Himmel, et al., "A Study of Manned Nuclear-Rocket Missions to Mars," p. 35.

10. *Ibid.*, p. 24.

11. *Ibid.*, p. 30.

12. *Ibid.*, p. 33-34.

13. John Logsdon, *The Decision to Go to the Moon: Project Apollo and the National Interest* (Cambridge, MA: MIT Press, 1970), pp. 111-12.

14. Office of Program Planning and Evaluation, "The Long Range Plan of the National Aeronautics and Space Administration," 16 December 1959, Logsdon, gen. ed., *Exploring the Unknown*, Vol. I, p. 404.

15. Ezell, "Man on Mars," p. 8.

16. Ernst Stuhlinger, "Possibilities of Electrical Space Ship Propulsion," Friedrich Hecht, editor, *Bericht über de V Internationalen Astronautischen Kongress* (Osterreichen Gesellschaft für Weltraumforschung, Vienna, Austria, 1955).

17. "Mars and Beyond," *The Wonderful World of Disney* television program, 4 December 1957.

18. Portree, *NASA's Origins*, p. 12.

19. Ernst Stuhlinger and Joseph King, "Concept for a Manned Mars Expedition with Electrically Propelled Vehicles," *Progress in Astronautics*, Vol. 9 (San Diego: Univelt, Inc., 1963), pp. 647-64.

20. *Ibid.*, p. 658.

21. *Ibid.*, p. 648.

22.  James Hansen, *Enchanted Rendezvous: John C. Houbolt and the Genesis of the Lunar-Orbit Rendezvous Concept* (Washington, DC: NASA Monographs in Aerospace History #4, 1995).

## Chapter 3

1.  Robert Sohn, "Summary of Manned Mars Mission Study," *Proceeding of the Symposium on Manned Planetary Missions: 1963/1964 Status* (Mountain View, CA: NASA TM X-53049, 1964), p. 151.

2.  T. A. Heppenheimer, *The Space Shuttle Decision: NASA's Search for a Reusable Space Vehicle* (Washington, DC: NASA, 1999), pp. 60-61.

3.  "One-Year Exploration-Trip Earth-Mars-Venus-Earth," Gaetano A. Crocco, Rendiconti del VII Congresso Internanzionale Astronautico, Associazione Italiana Razzi (paper presented at the Seventh Congress of the International Astronautical Federation, Rome, Italy, 1956), pp. 227-252.

4.  *Ibid.*, p. 239.

5.  Maxime Faget and Paul Purser, "From Mercury to Mars," *Aeronautics & Aerospace Engineering* (February 1963): 27.

6.  *Ibid.*, p. 24.

7.  Aeronutronic Division, Ford Motor Company, *EMPIRE, A Study of Early Manned Interplanetary Expeditions* (Huntsville, AL: NASA CR-51709, 21 December 1962).

8.  Lockheed Missiles & Space Company, *Manned Interplanetary Mission Study* (Lockheed Missiles and Space Company, March 1963).

9.  General Dynamics Astronautics, *A Study of Early Manned Interplanetary Missions Final Summary Report* (San Diego, CA, General Dynamics Astronautics, 31 January 1963).

10.  Aeronutronic, p. 1-2.

11.  Lockheed, p. xx.

12.  General Dynamics, p. 8-2.

13.  *Ibid.*, pp. 8-92 - 8-122.

14.  *Ibid.*, pp. 8-119 - 8-122.

15.  David Hammock and Bruce Jackson, "Vehicle Design for Mars Landing and Return to Mars Orbit," George Morgenthaler, editor, *Exploration of Mars* (San Diego, CA: Univelt, Inc., 1964), pp. 174-95.

16.  Raymond Watts, "Manned Exploration of Mars?" *Sky & Telescope* (August 1963): 63-67, 84.

17.  Hammock and Jackson, "Vehicle Design for Mars Landing," p. 175.

# Endnotes

18. Franklin Dixon, "Summary Presentation: Study of a Manned Mars Excursion Module," *Proceeding of the Symposium on Manned Planetary Missions: 1963/1964 Status* (Huntsville, AL: NASA TM X-53140, 1964), pp. 443-523.

19. *Ibid.*, p. 449.

20. *Ibid.*, p. 479.

21. *Ibid.*, p. 449.

22. *Ibid.*, p. 479.

23. J. N. Smith, *Manned Mars Missions in the Unfavorable (1975-1985) Time Period: Executive Summary Report* (Huntsville, AL: NASA TM X-53140, 1964).

24. *Ibid.*, p. 7.

25. *Ibid.*, pp. 11-12.

26. Sohn, "Summary of Manned Mars Mission Study," pp. 149-219.

27. *Ibid.*, p. 156.

28. *Ibid.*, p. 170.

29. *Ibid.*, p. 165-166.

30. "Part 17: Panel Discussion," *Proceeding of the Symposium on Manned Planetary Missions: 1963/1964 Status* (Huntsville, AL: NASA TM X-53140, 1964), pp. 748-749.

31. *Ibid.*, p. 751.

32. *Ibid.*

33. "Future Efforts to Stress Apollo Hardware," *Aviation Week & Space Technology* (16 November 1964): 48.

34. Ezell, "Man on Mars," p. 13.

35. *Ibid.*, p. 12.

36. Harry Ruppe, *Manned Planetary Reconnaissance Mission Study: Venus/Mars Flyby* (Huntsville, AL: NASA TM X-53205, 1965).

37. *Ibid.*, p. 53.

38. *Ibid.*, p. 7.

39. *Ibid.*, p. 8.

## Chapter 4

1. Robert Hotz, "New Era for NASA," *Aviation Week & Space Technology* (7 August 1967): 17.

2. Samuel Glasstone, *The Book of Mars* (Washington, DC: NASA SP-179, 1968), pp. 76-91.

3. William Hartmann and Odell Raper, *The New Mars: The Discoveries of Mariner 9* (Washington, DC: NASA SP-337, 1974), pp. 6-11.

4. Edward Clinton Ezell and Linda Neumann Ezell, *On Mars: Exploration of the Red Planet, 1958-1978* (Washington, DC: NASA 1984), pp. 74-82.

5. NASA, "A Report from Mariner IV," *NASA Facts* 3 (1966): 1.

6. *Ibid.*, pp. 5-6; Oran Nicks, *Summary of Mariner 4 Results* (Washington, DC: NASA SP-130), p. 35.

7. Hal Taylor, "LBJ Wants Post-Apollo Plans," *Missiles and Rockets* (4 May 1964); NASA, *Summary Report: Future Programs Task Group*, January 1965, Logsdon, gen. ed., *Exploring the Unknown*, Vol. I, p. 473.

8. "Future Effort to Stress Apollo Hardware," *Aviation Week & Space Technology* (16 November 1964): 48-51.

9. "Scientists Urge Priority for Mars Missions," *Aviation Week & Space Technology* (23 November 1964): 26.

10. Merrifield, "A Historical Note," p. 12; *Astronautics and Aeronautics 1966* (Washington, DC: NASA SP-4007), p. 17.

11. Willard Wilks and Rex Pay, "Quest for Martian Life Re-Emphasized," *Technology Week* (6 June 1966): 26-28.

12. Ezell and Ezell, *On Mars*, pp. 102-05.

13. Associate Administrator, Office of Space Science and Applications to Director, Office of Space Science and Applications, "Manned Planetary Missions Planning Group," 30 April 1965.

14. Franklin Dixon, "Manned Planetary Mission Studies from 1962 to 1968," IAA-89-729 (paper presented at the 40th Congress of the International Astronautical Federation, Malaga, Spain, 7-12 October 1989), p. 9.

15. Ezell, "Man on Mars," p. 12.

16. Merrifield, "A Historical Note," p. 13.

17. Planetary JAG, *Planetary Exploration Utilizing a Manned Flight System* (Washington, DC: NASA, 1966).

18. For example, see Robert Sohn, "A Chance for an Early Manned Mars Mission," *Astronautics & Aeronautics* (May 1965): 28-33.

19. Chief, NASA Kennedy Space Center Advanced Programs Office to Distribution, "Minutes of Joint Action Group Meeting of June 29-30, 1966," 8 July 1966.

# Endnotes

20.  R. R. Titus, "FLEM—Flyby-Landing Excursion Mode," AIAA Paper No. 66-36 (paper presented at the 3rd AIAA Aerospace Sciences Meeting, New York, New York, 24-26 January 1966).

21.  Edward Gray to H. K. Weidner, F. L. Williams, M. Faget, W. E. Stoney, J. West, J. P. Claybourne, and R. Hock, TWX, "Meeting to Establish Follow-on Activities Covering the Advanced Manned Planetary, Earth Orbital, and Lunar Exploration Programs," 17 November 1966.

22.  Edward Gray to H. K. Weidner, F. L. Williams, J. W. Carter, R. J. Harris, J. P. Claybourne, R. Hock, and R. J. Cerrato, TWX, "Follow-on Activity for Manned Planetary Program," 2 December 1966.

23.  John Logsdon, "From Apollo to the Space Shuttle: U.S. Space Policy, 1969-1972," unpublished manuscript, p. I-43.

24.  "U.S. Space Funding to Grow Moderately," *Aviation Week & Space Technology* (6 March 1967): 126.

25.  William Normyle, "Post-Apollo Program Potential Emerging," *Aviation Week & Space Technology* (6 March 1967): 126.

26.  President's Science Advisory Committee, *The Space Program in the Post-Apollo Period* (Washington, DC: The White House, February 1967).

27.  *Ibid.*, p. 18.

28.  "Science Advisors Urge Balanced Program," *Aviation Week & Space Technology* (6 March 1967): 135.

29.  William Normyle, "Manned Mars Flights Studied for the 1970s," *Aviation Week & Space Technology* (27 March 1967): 63.

30.  Merrifield, "A Historical Note," p. 13.

31.  Normyle, "Manned Mars Flights Studied," p. 62-63; Edward Gray and Franklin Dixon, "Manned Expeditions to Mars and Venus," Eric Burgess, editor, *Voyage to the Planets* (San Diego, CA: Univelt, Inc., 1967), pp. 107-35.

32.  "U.S. Space Funding Set to Grow Moderately," pp. 123-24.

33.  "House Unit Trims NASA Budget, Fight Pledged for Further Slashes," *Aviation Week & Space Technology* (22 May 1967): 24.

34.  "Space Funds Cut Deeply by House, Senate," *Aviation Week & Space Technology* (3 July 1967): 28.

35.  "Conferees Vote Space Cut," *Aviation Week & Space Technology* (7 August 1967): 24.

36.  William Normyle, "Small Hope Seen to Restore Space Funds," *Aviation Week & Space Technology* (10 July 1967): 38.

37.  Katherine Johnsen, "Webb Refuses to Choose Program for Cuts," *Aviation Week & Space Technology* (31 July 1967): 20.

38. Hotz, "New Era for NASA," p. 17.

39. Spacecraft Engineering Branch, *Apollo-based Venus/Mars Flybys* (Houston: NASA MSC, September 1967).

40. Contracting Officer to Prospective Contractors, "Planetary Surface Sample Return Probe Study for Manned Mars/Venus Reconnaissance/Retrieval Missions," Request for Proposal No. BG721-28-7-528P, 3 August 1967.

41. Irving Stone, "Manned Planetary Vehicle Study Proposed," *Aviation Week & Space Technology* (2 October 1967): 87.

42. William Normyle, "Priority Shift Blocks Space Plans," *Aviation Week & Space Technology* (11 September 1967): 27.

43. Ezell and Ezell, *On Mars*, p. 118.

44. "White House Stand Blocks NASA Budget Restoration," *Aviation Week & Space Technology* (28 August 1967): 32.

45. Ezell and Ezell, *On Mars*, p. 142.

## Chapter 5

1. NASA, "Outline of NASA Presentation to Space Task Group, August 4, 1969" (28 July 1969), p. 20.

2. Wernher von Braun, "The Next 20 Years of Interplanetary Exploration," *Astronautics & Aeronautics* (November 1965): 24.

3. NASA, *Astronautics and Aeronautics 1967* (Washington, DC: NASA SP-4008), pp. 339-41.

4. James Dewar, "Atomic Energy: The Rosetta Stone of Space Flight," *Journal of the British Interplanetary Society* (May 1994): 200.

5. *Ibid.*, p. 202.

6. John Kennedy, "Excerpts from 'Urgent National Needs,'" Logsdon, gen. ed., *Exploring the Unknown*, Vol. I, p. 454.

7. Dewar, "Atomic Energy," p. 203-04.

8. Raymond Watts, "Manned Exploration of Mars?" *Sky & Telescope* (August 1963): 64.

9. William House, "The Development of Nuclear Rocket Propulsion in the United States," *Journal of the British Interplanetary Society* 19, No. 8 (March-April 1964): 317-18.

10. Boeing Aerospace Group, *Integrated Manned Interplanetary Spacecraft Concept Definition, Vol. 1, Summary* (Seattle, Washington: NASA CR-66558, January 1968).

# Endnotes

11. North American Rockwell Corporation Space Division, *Definition of Experimental Tests for a Manned Mars Excursion Module: Final Report, Vol. 1, Summary* (SD 67-755-1, 12 January 1968).

12. Arthur Hill, "Apollo Shape Dominates NAR Manned Mars Study," *Aerospace Technology* (6 May 1968): pp. 26.

13. "Cost of Tet," *Aviation Week & Space Technology* (27 May 1968): 25.

14. "Congressional Critics Aim to Cut NASA Budget to $4-Billion Level," *Aviation Week & Space Technology* (12 February 1968): 22.

15. Katherine Johnsen, "NASA Gears for $4-Billion Fund Limit," *Aviation Week & Space Technology* (27 May 1968): 30.

16. "Webb Urges Full $4-Billion NASA Fund," *Aviation Week & Space Technology* (1 July 1968): 22.

17. Administrator to Associate Administrator for Manned Space Flight, "Termination of the Contract for Procurement of Long Lead Time Items for Vehicles 516 and 517," Logsdon, gen. ed., *Exploring the Unknown*, Vol. I, pp. 494-95.

18. "Work on Future Saturn Launchers Halted," *Aviation Week & Space Technology* (12 August 1968): 30.

19. NASA, *Astronautics and Aeronautics 1968* (Washington, DC: NASA SP-4010), pp. 212-13.

20. NASA, *Astronautics and Aeronautics 1968* (Washington, DC: NASA SP-4010), p. 215.

21. William Normyle, "NASA Plans Five-Year Fund Rise," *Aviation Week & Space Technology* (14 October 1968): 16.

22. Bureau of the Budget, "National Aeronautics and Space Administration: Highlight Summary," 30 October 1968, Logsdon, gen. ed., *Exploring the Unknown*, Vol. I, pp. 497-98.

23. Courtney Brooks, James Grimwood, and Loyd Swenson, Jr., *Chariots for Apollo, A History of Manned Lunar Spacecraft* (Washington, DC: NASA SP-4205, 1979), p. 279.

24. *Ibid.*, pp. 256-60.

25. Arthur Schlesinger, Jr., *The Almanac of American History* (Greenwich, CT: Brompton Books, 1993), p. 581.

26. "Against the Tide," *Aviation Week & Space Technology* (17 March 1969): 15.

27. Heppenheimer, *The Space Shuttle Decision*, pp. 115-16.

28. Roger Launius, "The Waning of the Technocratic Faith: NASA and the Politics of the Space Shuttle Decision," Philippe Jung, editor, *History of Rocketry and Astronautics*, AAS History Series, Volume 21 (San Diego, CA: Univelt, Inc., 1997), p. 190.

29. Heppenheimer, *The Space Shuttle Decision*, p. 127.

30. NASA, *Astronautics and Aeronautics 1968* (Washington, DC: NASA SP-4010), pp. 215.

31. Charles Townes, et al., "Report of the Task Force on Space," 8 January 1969, Logsdon, gen. ed., *Exploring the Unknown*, Vol. I, p. 502.

32. *Ibid.*, p. 505.

33. Heppenheimer, *The Space Shuttle Decision*, pp. 121-22.

34. Dwayne Day, "Viewpoint: Paradigm Lost," *Space Policy* (August 1995): 156.

35. Heppenheimer, *The Space Shuttle Decision*, pp. 127-28.

36. William Normyle, "NASA Aims at 100-man Station," *Aviation Week & Space Technology* (24 February 1969): 16.

37. Richard Nixon, "Memorandum for the Vice President, the Secretary of Defense, the Acting Administrator, NASA, and the Science Advisor," 13 February 1969, Logsdon, gen. ed., *Exploring the Unknown*, Vol. I, p. 513.

38. Thomas Paine, "Problems and Opportunities in Manned Space Flight," Logsdon, gen. ed., *Exploring the Unknown*, Vol. I, pp. 513-19.

39. Logsdon, "From Apollo to the Space Shuttle," pp. III-7 - III-8; Heppenheimer, *The Space Shuttle Decision*, pp. 130-31.

40. NASA, "Integrated Manned Space Flight Program, 1970-1980" (12 May 1969).

41. *Ibid.*, p. 2.

42. Logsdon, "From Apollo to the Space Shuttle," p. IV-50.

43. Logsdon, "From Apollo to the Space Shuttle," p. IV-40.

44. NASA, *Astronautics and Aeronautics 1969* (Washington, DC: NASA SP-4014), pp. 235-36.

45. NASA, *Astronautics and Aeronautics 1969* (Washington, DC: NASA SP-4014), p. 239.

46. "Washington Roundup," *Aviation Week & Space Technology* (21 July 1969): 15.

47. NASA, *Astronautics and Aeronautics 1969* (Washington, DC: NASA SP-4014), p. 270.

48. NASA, *Astronautics and Aeronautics 1969* (Washington, DC: NASA SP-4014), p. 271.

49. NASA, "Outline of NASA Presentation to Space Task Group, August 4, 1969" (28 July 1969), p. 20.

50. Wernher von Braun, "Manned Mars Landing Presentation to the Space Task Group," presentation materials (4 August 1969).

51. *Ibid.*, p. 4.

# Endnotes

52. *Ibid.*, pp. 22-24.

53. *Ibid.*, p. 26.

54. *Ibid.*, p. 35.

55. *Ibid.*, pp. 41-43.

56. NASA, "Outline of NASA Presentation," p. 23.

57. Robert Seamans, Jr., Secretary of the Air Force, to Spiro Agnew, Vice President, letter, 4 August 1969, Logsdon, gen. ed., *Exploring the Unknown*, Vol. I, p. 521-22.

58. "Washington Roundup," p. 15.

59. Logsdon, "From Apollo to the Space Shuttle," p. IV-53.

60. *Ibid.*

61. *Ibid.*, pp. 57-63.

62. "Space Manpower," *Aviation Week & Space Technology* (11 August 1969): 25.

63. Robert Hotz, "The Endless Frontier," *Aviation Week & Space Technology* (11 August 1969): 17.

64. William Normyle, "Manned Mission to Mars Opposed," *Aviation Week & Space Technology* (18 August 1969): 16.

65. *Ibid.*, p. 17.

66. NASA, *America's Next Decades in Space: A Report to the Space Task Group* (Washington, DC: NASA, September 1969).

67. *Ibid.*, p. 7.

68. *Ibid.*, p. 1.

69. Space Task Group, *The Post-Apollo Space Program: Directions for the Future* (Washington, DC: NASA, September 1969).

70. Logsdon, gen. ed., *Exploring the Unknown*, Vol. I, pp. 522-23.

71. Space Task Group, *The Post-Apollo Space Program*, pp. ii-iii.

72. *Ibid.*, p. iv.

73. *Ibid.*

74. Wernher von Braun interview by John Logsdon, referenced in T. A. Heppenheimer, *The Space Shuttle Decision*, p. 152.

75.  Robert Mayo, Director, Bureau of the Budget, "Memorandum for the President, 'Space Task Group Report,'" 25 September 1969, Logsdon, gen. ed., *Exploring the Unknown*, Vol. I, pp. 545-46.

76.  "NASA Budget Faces House-Senate Parley," *Aviation Week & Space Technology* (29 September 1969): 19.

77.  George Mueller to John Naugle, 6 October 1969; Morris Jenkins, *Manned Exploration Requirements and Considerations* (Houston: NASA, February 1971), pp. iii-iv.

78.  Logsdon, "From Apollo to the Space Shuttle," p. V-22.

79.  "Bill of Fare," *Aviation Week & Space Technology* (2 February 1970): 11; NASA, *Astronautics and Aeronautics 1970* (Washington, DC: NASA SP-40), pp. 11-12.

80.  "Centers Reviewed," *Aviation Week & Space Technology* (19 January 1970): 16.

81.  "Space in the 1970s," *Aviation Week & Space Technology* (9 February 1970): 11.

82.  Schlesinger, p. 586.

83.  "Ad Astra per Aspera," *Aviation Week & Space Technology* (9 February 1970): 10.

84.  Logsdon, "From Apollo to the Space Shuttle," p. V-40.

85.  Space Science and Technology Panel of the President's Science Advisory Committee, *The Next Decade in Space* (Washington, DC: Executive Office of the President, Office of Science and Technology, March 1970), pp. 3, 22.

86.  *Ibid.*, p. i.

87.  *Ibid.*, p. 45.

88.  *Ibid.*, p. 4.

89.  *Ibid.*, p. 52.

90.  Launius, "The Waning of the Technocratic Faith," p. 185.

91.  Morris Jenkins, *Manned Mars Exploration Requirements and Considerations* (Houston: NASA, February 1971), p. iv.

92.  *Ibid.*, p. iii.

93.  *Ibid.*, p. 4–14.

94.  *Ibid.*, p. 2–15.

95.  U.S. Congress, *Nuclear Rocket Development Program*, Joint Hearings before the Committee on Aeronautical and Space Sciences, United States Senate and the Joint Committee on Atomic Energy, 92nd Congress of the United States, First Session, 23-24 February 1971, p. 1.

## Endnotes

96. *Ibid.*, pp. 13-15.

97. *Ibid.*, p. 21.

98. *Ibid.*, p. 34.

99. *Ibid.*, p. 40.

100. "OMB Limits NASA to $15 Million for NERVA," *Aviation Week & Space Technology* (4 October 1971): 20.

101. Launius, "The Waning of the Technocratic Faith," pp. 188-89.

102. NASA, *Astronautics and Aeronautics 1972*, (Washington, DC: NASA SP-4017), pp. 4-5

103. Dewar, "Atomic Energy," p. 205.

## Chapter 6

1. Benton Clark, "The Viking Results—The Case for Man on Mars ," AAS 78-156, Richard Johnston, Albert Naumann, and Clay Fulcher, editors, *The Future U.S. Space Program* (San Diego: Univelt, Inc., 1978), p. 263.

2. Hartmann and Raper, *The New Mars*, pp. 32.

3. Ray Bradbury, Arthur C. Clarke, Bruce Murray, Carl Sagan, and Walter Sullivan, *Mars and the Mind of Man* (New York: Harper and Row, 1973) is an informative and entertaining exploration of changing human perceptions of the planet Mars.

4. Hartmann and Raper, *The New Mars*, pp. 94-107.

5. Andrew Wilson, *Solar System Log* (New York: Jane's, 1987), p. 69.

6. Richard Lewis, "On the Golden Plains of Mars," *Spaceflight* (October 1976): 364.

7. Richard Lewis, "The Puzzle of Martian Soil, " *Spaceflight* (November 1976): 391-95. See also Bevan French, *Mars: The Viking Discoveries* (Washington, DC: NASA, 1977), pp. 20-22; Andrew Chaikin, "The Case for Life on Mars," *Air & Space Smithsonian* (February/March 1991): 63-71; Harold Klein, Norman Horowitz, and Klaus Biemann, "The Search for Extant Life on Mars ," Hugh Kieffer, Bruce Jakosky, Conway Snyder, and Mildred Mathews, editors, *Mars*, (Tucson: University of Arizona Press, 1992), pp. 1221-33.

8. Cary Spitzer, editor, *Viking Orbiter Views of Mars* (Washington, DC: NASA, 1980), pp. 31-32. See also Victor Baker, Michael Carr, Virginia Gullick, Cameron Williams, and Mark Marley , "Channels and Valley Networks"; and Christopher McKay, R. L. Mancinelli, Carol Stoker, and R. A. Wharton, "The Possibility of Life on Mars During a Water-Rich Past," both in Hugh Kieffer, et al., editors, *Mars*, pp. 493-522 and 1234-45.

9. Ezell and Ezell, *On Mars*, pp. 422-23.

10. NASA, *Proceedings of the Seventh Annual Working Group on Extraterrestrial Resources* (Washington, DC: NASA SP-229, 1970), p. iii.

11. J. N. Smith, *Manned Mars Missions in the Unfavorable (1975-1985) Time Period*, pp. 11-12.

12. Louis Friedman interview by David S. F. Portree, 15 August 1999.

13. R. L. Ash, W. L. Dowler, and G. Varsi, "Feasibility of Rocket Propellant Production on Mars," *Acta Astronautica* (July-August 1978): 705-24.

14. Clark, "The Viking Results," p. 273.

15. *Ibid.*, p. 274.

16. Other examples of Mars ISRU papers in the 1980s include the following: Benton Clark, "The Chemistry of the Martian Surface: Resources for the Manned Exploration of Mars ," AAS 81-243, Penelope Boston, editor, *The Case for Mars*, (San Diego, CA: Univelt, Inc., 1984), pp. 197-208; G. R. Babb and W. R. Stump, "The Effect of Mars Surface and Phobos Propellant Production on Earth Launc h Mass," Michael Duke and Paul Keaton, editors, *Manned Mars Missions: Working Group Papers*, Vol. 1 (Huntsville, AL, and Los Alamos, NM: NASA M002, NASA/LANL, June 1986), pp. 162-175; R. H. Frisbee, "Mass and Power Estimates for Mars In-Situ Propellant Production Systems," AIAA-87-1900 (papers presented at the AIAA/SAE/ASME/ASEE 23rd Joint Propulsion Conference , 29 June-2 July 1987); Benton Clark and Donald P ettit, "The Hydrogen Peroxide Economy on Mars ," AAS 87-214, Carol Stoker, editor, *The Case for Mars III: Strategies for Exploration—General Interest and Overview* (San Diego, CA: Univelt, Inc.,1989), pp. 551-57; Robert Ash, Joseph Werne, and Merry Beth Haywood, "Design of a Mars Oxygen Processor," AAS 87-263, Carol Stoker, editor, *The Case for Mars III: Strategies for Exploration—Technical* (San Diego, CA: Univelt, Inc.,1989), pp. 479-87; Diane L. Galecki, "In-Situ Propellant Advantages for Fast Transfer to Mars," AIAA-88-2901 (paper presented at the AIAA/ASME/SAE/ASEE 24th J oint Propulsion Conference , 11-13 July 1988); Thomas Meyer and Christopher McKa y, "The Resources of Mars for Human Settlement, " *Journal of the British Interplanetary Society* (April 1989): 147-60; J. R. French, "Rocket Propellants from Martian Resources ," *Journal of the British Interplanetary Society* (April 1989): 167-70.

17. Louis Friedman interview, 15 August 1999. Friedman founded The Planetary Society with Carl Sagan and Bruce Murray in 1980.

18. Robert Ash interview by David S. F. Portree, 29 July 1999. Ash called Friedman "ISRU's godparent."

## Chapter 7

1. Alcestis Oberg, "The Grass Roots of the Mars Conference ," AAS 81-225, Penelope Boston, editor, *The Case for Mars* (San Diego, CA: Univelt, Inc., 1984), p. ix.

2. Tim Furniss, *Space Shuttle Log* (New York: Jane's, 1986), pp. 15-18, 34-36.

3. William Stockton and John Noble Wilford, *Spaceliner* (New York: Times Books, 1981), p. 159.

4. Oberg, "The Grass Roots," p. ix.

5. Benton Clark interview by David S. F. Portree, 27 August 1999.

6. S. Fred Singer, "The PH-D Proposal: A Manned Mission to Phobos and Deimos ," AAS 81-231, Penelope Boston, editor, *The Case for Mars* (San Diego, CA: Univelt, Inc., 1984), pp. 39-65.

# Endnotes

7.   S. Fred Singer, "To Mars By Way of Its Moons," *Scientific American* (March 2000): 56-57.

8.   Space Sciences Department, *Manned Lunar, Asteroid and Mars Missions, Visions of Space Flight: Circa 2001* (Schaumburg, IL: Science Applications International Corporation, September 1984).

9.   Louis Friedman, "Visions of 2010," *The Planetary Report* (March/April 1985): 5.

10.  Louis Friedman interview by David S. F. Portree, 15 August 1999.

11.  Friedman, "Visions," pp. 6, 22.

12.  *Ibid.*, p. 22.

13.  "Beggs Calls for Start on Space Station," *Space News Roundup* (25 June 1982): 1, 3-4.

14.  Clarke Covington, "The Role of the Space Operations Center," presentation materials (28 May 1981).

15.  Dave Alter, "Space Operations Center" (Houston, TX: NASA Johnson Space Center Press Release 82-008, 19 February 1982).

16.  *Presidential Papers of the President: Administration of Ronald Reagan, 1985* (Washington, DC: U.S. Government Printing Office, 1985), p. 90.

17.  Humboldt Mandell, personal communication.

18.  Michael Duke interview by David S. F. Portree, 26 August 1999.

19.  Paul Keaton interview by David. S. F. Portree, 30 August 1999.

20.  R. F. Baillie to R. W. Johnson, "Manned Planetary Exploration Action Item from the Wallops Workshop" (August 1, 1978); Joseph Loftus, Jr., interview by David S. F. Portree, 15 August 1999.

21.  Keaton interview, 30 August 1999.

22.  Harrison Schmitt, "A Millennium Project—Mars 2000," Wendell Mendell, editor, *Lunar Bases and Space Activities of the 21st Century* (Houston, TX: Lunar and Planetary Science Institute, 1985), p. 787.

23.  Duke interview, 30 August 1999; Keaton interview, 30 August 1999.

24.  *Ibid.*

25.  Michael Duke and P aul Keaton, editors, *Manned Mars Missions, Working Group Summary Report* (Huntsville, AL, and Los Alamos, NM: NASA M001, NASA/LANL, May 1986); Michael Duke and P aul Keaton, editors, *Manned Mars Missions, Working Group Papers*, Vol. 1 and Vol. 2 (Huntsville, AL, and Los Alamos, NM: NASA M002, NASA/LANL, June 1986).

26.  Charles Cravotta and Melanie DeForth, "Soviet Plans for a Manned Flight to Mars" (Office of Scientific and Weapons Research, U.S. Central Intelligence Agency, 2 April 1985), p. 2.

27.   *Ibid.*

28.   *Ibid.*, p. 7.

29.   *Ibid.*, p. 8.

30.   Barney Roberts, "Concept for a Manned Mars Flyby," *Manned Mars Missions: Working Group Papers*, Vol. 1 (Huntsville, AL, and Los Alamos, NM: NASA M002, NASA/LANL, June 1986), pp. 203-18.

31.   *Ibid.*, pp. 213-15.

32.   Buzz Aldrin, "The Mars Transit System," *Air & Space Smithsonian* (October/November 1990): 47.

33.   Charles Rall and Walter Hollister, "Free-fall Periodic Orbits Connecting Earth and Mars," AIAA No. 71-92 (paper presented at the American Institute of Aeronautics and Astronautics 9th Aerospace Sciences Meeting, New York, New York, 25-27 January 1971). Cycler proponent Buzz Aldrin was one of Hollister's students at MIT before he became a NASA astronaut.

34.   S. M. Welch and C. R. Stoker, editors, *The Case for Mars: Concept Development for a Mars Research Station* (Boulder, CO: Boulder Center for Science Policy, 10 April 1986).

35.   Thomas Paine, "A Timeline for Martian Pioneers," AAS 84-150, Christopher McKay, editor, *The Case for Mars II* (San Diego, CA: Univelt, Inc., 1985), pp. 18-19.

36.   Michael Duke, Wendell Mendell, and Barney Roberts, "Lunar Base: A Stepping Stone to Mars," AAS 84-162, Christopher McKay, editor, *The Case for Mars II* (San Diego, CA: Univelt, Inc., 1985), pp. 207-20.

37.   Humboldt Mandell, "Space Station—The First Step," AAS 84-160, Christopher McKay, editor, *The Case for Mars II* (San Diego, CA: Univelt, Inc.,1985), pp. 157-70.

38.   Welch and Stoker, *The Case for Mars: Concept Development*, p. 53.

39.   For examples, see Robert Farquhar, "Lunar Communications with Libration-Point Satellites," *Journal of Spacecraft and Rockets* (October 1967): 1383, and Robert Farquhar, "A Halo-Orbit Lunar Station, " *Astronautics & Aeronautics* (June 1972): 59-63. In 1971, Farquhar became involved in Harrison Schmitt's effort to target Apollo 17 to the lunar farside crater Tsiolkovskii. He studied the possibility of placing communication relay satellites in Lagrange point halo orbits to permit continuous communication between the Apollo 17 moonwalkers at Tsiolkovskii and Mission Control on Earth ("Lunar Backside Landing for Apollo 17," presentation materials, 2 September 1971).

40.   Robert Farquhar and David Dunham, "Libration-Point Staging Concepts for Earth-Mars Transportation," *Manned Mars Missions: Working Group Papers*, Vol. 1 (Huntsville, AL, and Los Alamos, NM: NASA M002, NASA/LANL, June 1986), pp. 66-77.

41.   Paul Keaton, *A Moon Base/Mars Base Transportation Depot* (Los Alamos, NM: LA-10552-MS, UC-34B, Los Alamos National Laboratory, September 1985).

42.   *Ibid.*, p. 10.

# Endnotes

## Chapter 8

1.  [Carl Sagan, Louis Friedman, and Bruce Murra y], The Mars Declaration, special supplement to *The Planetary Report* (November/December 1987). Author names revealed in Louis F riedman interview, 15 August 1999.

2.  National Commission on Space (NCOS), *Pioneering the Space F rontier: The Report of the National Commission on Space* (New York: Bantam Books, May 1986).

3.  Lyn Ragsdale, "Politics Not Science: The U.S. Space Program in the Reagan and Bush Years," *Spaceflight and the Myth of Presidential Leader ship*, Roger Launius and Howard McCurdy, editors (Urbana, IL: University of Illinois Press, 1997), p. 151.

4.  Carole Shifrin, "NASA Nears Final Decisions on Station Configuration," *Aviation Week & Space Technology* (10 March 1986): 107-109; NASA, "NASA Facts: Space Station" (Kennedy Space Center Press Release No . 16-86, January 1986).

5.  "NASA Managers Divided on Station, " *Aviation Week & Space Technology* (28 July 1986): 24-25; Craig Covault, "Launch Capacity, EVA Concerns F orce Space Station Re-Design, " *Aviation Week & Space Technology* (21 July 1986): 20; NASA, Space Station Freedom Media Handbook (Washington, DC: NASA, April 1989), p. 7; Mark Hess, "NASA Proceeding Toward Space Station Development" (Johnson Space Center Press Release 87-50, 3 April 1987), p. 2.

6.  Paul Mann, "Commission Sets Goals for Moon, Mars Settlement in 21st Century," *Aviation Week & Space Technology* (24 March 1986): 18-21.

7.  Thomas Paine, "Overview: Report of the National Commission on Space," Duke Reiber, editor, *The NASA Mars Conference* (San Diego: Univelt, Inc., 1988), p. 533.

8.  NCOS, *Pioneering*, p. 191.

9.  "Spaced Out," *Aviation Week & Space Technology* (15 September 1986): 11.

10. Thomas Paine, "Who Will Lead the World's Next Age of Discovery?" *Aviation Week & Space Technology* (21 September 1987): 43.

11. Craig Covault, "Ride Panel Calls for Aggressive Action to Assert U.S. Leadership in Space," *Aviation Week & Space Technology* (24 August 1987): 26.

12. NASA, "Statement by Dr . Sally K. Ride, Associate Administrator for Exploration (Acting) before the Subcommittee on Space Science and Applications, Committee on Science, Space, and Technology, House of Representatives" (22 July 1987), p. 1.

13. Sally Ride, *Leadership and America's Future in Space* (Washington, DC: NASA, August 1987), p. 5.

14. Ride, *Leadership*, p. 21.

15. NASA, "Statement by Dr. Sally K. Ride," p. 2.

16. Ride, *Leadership*, p. 53.

17. *Ibid.*, p. 6.

18. Craig Covault, "Ride Panel Calls for Aggressive Action," p. 26; NASA, *Astronautics and Aeronautics 1986-1990* (Washington, DC: NASA SP-4027), p. 126.

19. Ride, *Leadership*, p. 55.

20. Craig Covault, "Ride Panel Will Urge Lunar Base , Earth Science as New Space Goals ," *Aviation Week & Space Technology* (13 July 1987): 17; see also Michael Collins, *Mission to Mars* (New York: Grove Weidenfeld, 1990), pp. xii, 197.

21. Ride, *Leadership*, p. 55.

22. *Ibid.*, p. 22.

23. *Ibid.*, p. 43.

24. NASA, "Statement by Dr. Sally K. Ride," p. 4.

25. Ride, *Leadership*, p. 40.

26. "NASA Forms Office to Study Manned Lunar Base , Mars Missions," *Aviation Week & Space Technology* (8 June 1987): 22.

27. Ride, *Leadership*, p. 53.

28. *Ibid.*, p. 47.

29. Science Applications International Corporation, *Piloted Sprint Missions to Mars* (Schaumberg, IL: Report No. SAIC-87/1908, Study No. 1-120-449-M26, November 1987).

30. *Ibid.*, p. 2.

31. *Ibid.*, p. 13; University of Texas and Texas A&M University Design Team, "To Mars—A Manned Mars Mission Study," Summer Project Report (NASA Universities Advanced Space Design Program, Advanced Programs Office, Johnson Space Center, August 1985).

32. *Ibid.*, p. 17.

33. NASA, *Astronautics and Aeronautics 1986-1990*, p. 115.

34. Office of Exploration, *Exploration Studies Technical Report, FY 1988 Status, Volume 1: Technical Summary* (Washington, DC: NASA TM-4075, December 1988); Office of Exploration, "FY88 Exploration Studies Technical Presentation to the Administrator," presentation materials (25 July 1988).

35. Martin Marietta, *Manned Mars System Study (MMSS) Executive Summary* (Denver, CO: Martin Marietta, July 1990).

# Endnotes

36.   Clark interview, 27 August 1999.

37.   David S. F. Portree, *Thirty Years Together: A Chronology of U.S.-Soviet Space Cooperation* (Houston: NASA CR-185707, February 1993), pp. 26-27.

38.   Harvey Meyerson, "Spark Matsunaga 1916-1990," *The Planetary Report* (July/August 1990): 26.

39.   Philip Klass, "Commission Considers Joint Mars Exploration, Lunar Base Options," *Aviation Week & Space Technology* (29 July 1985): 47.

40.   Carl Sagan, "To Mars," *Aviation Week & Space Technology* (8 December 1986): 10.

41.   The Mars Declaration.

42.   Richard O'Lone, "Scientist Sees Space Station Useful Only If Linked to Manned Mars Mission, " *Aviation Week & Space Technology* (25 January 1988): 55, 57.

43.   Portree, *Thirty Years Together*, p. 30.

44.   V. Glushko, Y. Semyonov, and L. Gorshkov, "The Way to Mars," *The Planetary Report* (November-December 1988): 4-8. Translation of *Pravda* article dated 24 May 1988.

## Chapter 9

1.    NASA, *Report of the 90-Day Study on Human Exploration of the Moon and Mar s* (Washington, DC: NASA, November 1989), pp. 9-12 - 9-13.

2.    Aaron Cohen interview by David S. F. Portree, 27 August 1999.

3.    Craig Covault, "Space Policy Outlines Program to Regain U .S. Leadership," *Aviation Week & Space Technology* (22 February 1988): 20.

4.    "NASA Funds $100-Million Pathfinder Program for Mars, Lunar Technology," *Aviation Week & Space Technology* (18 January 1988): 17.

5.    Dwayne Day, "Doomed to Fail," *Spaceflight* (March 1995): 80.

6.    Office of the White House Press Secretary, "Remarks of the President at the 20th Anniversary of Apollo Moon Landing" (Washington, DC: White House, 20 July 1989).

7.    "Space Wraith," *Aviation Week & Space Technology* (24 July 1989): 21.

8.    Mark Craig interview by David S. F. Portree, 13 September 1999.

9.    Richard Truly and Franklin Martin, "Briefing to NASA Employees," presentation materials (26 July 1989).

10.   Ivan Bekey interview by David S. F. Portree, 7 September 1999.

11.  "NASA Accelerates Lunar Base Planning as Station Changes Draw European Fire," *Aviation Week & Space Technology* (18 September 1999): 26-27.

12.  Humboldt Mandell interview by David S. F. Portree, 13 September 1999.

13.  Cohen interview, 27 August 1999.

14.  *Ibid*.

15.  NASA, "Cost Summary," unpublished chapter in *Report of the 90-Day Study on Human Exploration of the Moon and Mars*, p. 2.

16.  *Ibid*., p. 3.

17.  *Ibid*.

18.  *Ibid*., p. 4.

19.  Ivan Bekey, "A Smaller Scale Manned Mars Evolutionary Program, " IAF-89-494 (paper presented at the 40th Congress of the International Astronautical Federation, Malaga, Spain, 7-12 October 1989), p. 6.

20.  Bekey interview, 7 September 1999.

21.  Rod Hyde, Yuki Ishikawa, and Lowell Wood, "An American-Traditional Space Exploration Program: Quick, Inexpensive, Daring, and Tenacious, Briefing to the National Space Council"(Livermore, CA: LLNL Doc. No. Phys. Brief 89-403, September 1989).

22.  Day, "Doomed to Fail," p. 81.

23.  "Space Policy," *Aviation Week & Space Technology* (30 October 1989): 15; John Connolly, personal communication.

24.  Craig interview, 13 September 1999.

25.  Roderick Hyde, Muriel Ishikawa, and Lowell Wood, "Mars in this Century: The Olympia Project," UCRL-98567, DE90 008356, Lawrence Livermore National Laboratory (paper presented at the U.S. Space Foundation 4th National Space Symposium, Colorado Springs, Colorado, 12-15 April 1988).

26.  R. A. Hyde, M. Y. Ishikawa, and L. L. Wood, "Toward a Permanent Lunar Settlement in the Coming Decade: The Columbus Project" (Lawrence Livermore National Laboratory: UCRL-93621, DE86 006709, 19 November 1985).

27.  Hyde, *et al.*, "An American-Traditional Space Exploration Program," p. 38.

28.  *Ibid*., p. 3-4.

29.  "Notice to NASA," *Aviation Week & Space Technology* (15 January 1990): 15.

30.  Committee on the Human Exploration of Space , *Human Exploration of Space: A Review of NASA's 90-Day Study and Alternatives* (Washington, DC: National Academy Press, 1990), p. x.

# Endnotes

31.  *Ibid.*, p. 28.

32.  *Ibid.*, p. 3.

33.  *Ibid.*, pp. xii-xiii.

34.  *Ibid.*, p. x.

35.  "Bush Calls for Two Proposals for Missions to Moon, Mars," *Aviation Week & Space Technology* (12 March 1990): 18.

36.  Breck Henderson, "Livermore Plan for Exploring Moon, Mars Draws Space Council Attention," *Aviation Week & Space Technology* (22 January 1990): 84.

37.  Douglas Isbell, "Congress Says OK to Moon, Mars Work," *Space News* (28 May-3 June 1990): 3, 20.

38.  Douglas Isbell, "Ex-Astronaut Stafford to Head Moon-Mars Outreach Team," *Space News* (4-10 June 1990): 4.

39.  Mandell interview, 13 September 1999.

40.  Andrew Lawler, "Bush: To Mars by 2019," *Space News* (14-20 May 1990): 1.

41.  Patricia Guilmartin, "House Kills Funding for Moon/Mars Effort," *Aviation Week & Space Technology* (2 July 1990): 28.

42.  "Darman Backs NASA," *Aviation Week & Space Technology* (21 May 1990): 17.

43.  Douglas Isbell and Andrew Lawler, "Senators Assail Bush Plan," *Space News* (7-13 May 1990): 1.

44.  "Bush Sets 2019 Manned Mars Objective," *Aviation Week & Space Technology* (21 May 1990): 19.

45.  Andrew Lawler, "Bush Moon-Mars Plan Handed First Defeat," *Space News* (18-24 June 1990): 3.

46.  NASA, *Astronautics and Aeronautics 1986-1990* (Washington, DC: NASA SP-4027), pp. 272-73.

47.  Craig Covault, "White House Endorses Plan for Shuttle , Station Scale-Back," *Aviation Week & Space Technology* (17 December 1990): 20; NASA, *Astronautics and Aeronautics 1986-1990*, p. 287.

48.  "U.S. Astronaut to Visit Soviet Station, Cosmonaut to Fly on Shuttle," *Aviation Week & Space Technology* (22 October 1990): 24.

49.  "Senior Soviet Space Officials Outline Plan for Joint Mars Mission," *Aviation Week & Space Technology* (19 November 1990): 67; Arnold Aldrich to Distribution, "Background Material on Cooperation with NPO Energia" (29 June 1992).

50.  Leonard David, "Faster, Cheaper Mars Exploration Proposed," *Space News* (11-17 June 1990): 4.

51.  Yuri Semyonov and Leonid Gorshkov, "Destination Mars," *Science in the USSR* (July-August 1990): 15-18.

52. *Ibid.*, p. 17.

53. Scientific Industrial Corporation "Energia," Mars Manned Mission: Scientific/Technical Report (Moscow, Russia: USSR Ministry of General Machinery, 1991), p. 1.

54. *Ibid.*, p. 15.

55. SEI Synthesis Group, *America at the Threshold: America's Space Exploration Initiative* (Washington, DC: Government Printing Office, May 1991).

56. "Reaching Out," *Aviation Week & Space Technology* (4 June 1990): 15; Craig Covault, "Exploration Initiative Work Quickens as Some Concepts Avoid Station," *Aviation Week & Space Technology* (17 September 1990): 36.

57. *Astronautics and Aeronautics 1986-1990*, p. 255.

58. Covault, "Exploration Initiative Work Quickens," p. 36.

59. *America at the Threshold*, p. 52.

60. *Ibid.*, p. 8.

61. Kent Joosten, personal communication.

## Chapter 10

1. Kent Joosten, Ryan Schaefer, and Stephen Hoffman, "Recent Evolution of the Mars Reference Mission, " AAS-97-617 (paper presented at the AAS/AIAA Astrodynamic Specialist Conference, Sun Valley, Idaho, 4-7 August 1997), p. 1.

2. Robert Zubrin with Richard Wagner, *The Case for Mars* (New York: Free Press, 1996), pp. 51-52; Benton Clark interview by David S. F. Portree, 30 September 1999.

3. Zubrin and Wagner, *The Case for Mars*, p. 65.

4. Leonard David, "Faster, Cheaper Mars Exploration," p. 37.

5. Robert Zubrin and David Baker, "Humans to Mars in 1999," *Aerospace America* (August 1990): 30-32, 41. For other examples, see Zubrin and Benjamin Adelman, "The Direct Route to Mars," Final Frontier (July/August 1992): 10-15, 53, 55; Zubrin and Christopher McKay, "Pioneering Mars," *Ad Astra* (September/October 1992): 34-41; Zubrin, "The Significance of the Martian Frontier," *Ad Astra* (September/October 1994): 30-37; Zubrin, "Mars: America's New Frontier," Final Frontier (May/June 1995): 42-46; Zubrin, "The Economic Viability of Mars Colonization," *Journal of the British Interplanetary Society* (October 1995): 407-414; Zubrin, "The Promise of Mars," *Ad Astra* (May/June 1996): 32-38; Zubrin, "Mars on a Shoestring," *Technology Review* (November/December 1996): 20-31; Zubrin, "Sending Humans to Mars," *Scientific American Presents* (Spring 1999): 46-51; Zubrin, "The Mars Direct Plan," *Scientific American* (March 2000): 52-55.

# Endnotes

6.  Zubrin and Baker, p. 30.

7.  Ibid, p. 31.

8.  Martin Marietta, *Manned Mars System Study (Mar s Transportation and Facility Infrastructure Study), Volume II, Final Report* (Denver, CO: Martin Marietta, July 1990), pp. 4-11 - 4-16.

9.  Zubrin and Baker, p. 41.

10. Michael Duke and Nancy Anne Budden, editors, *Mars Exploration Study Workshop II* (Houston: NASA CP-3243, November 1993), p. iii.

11. Exploration Programs Office, "EXPO Mars Program Study, Presentation to the Associate Administrator for Exploration," presentation materials (9 October 1992).

12. Zubrin with Wagner, *The Case for Mars*, pp. 66-67.

13. David Weaver and Michael Duke, "Mars Exploration Strategies: A Reference Program and Comparison of Alternative Architectures," AIAA 93-4212 (paper presented at the AIAA Space Program and Technologies Conference, Huntsville, Alabama, 21-23 September 1993).

14. Duke and Budden, *Mars Exploration Study Workshop II*.

15. Robert Zubrin and David Weaver, "Practical Methods for Near-Term Piloted Mars Missions," AIAA 93-2089 (paper presented at the AIAA/SAE/ASME/ASEE 29th Joint Propulsion Conference, Monterey, California, 28-30 June 1993), p. 3. In a 30 September 1999 interview with the author, Benton Clark compared the Mars Direct ERV volume per crewmember to "a telephone booth." See also David S. F. Portree, "The New Martian Chronicles," *Astronomy* (July 1997): 32-37.

16. Kent Joosten, personal communication.

17. Donald Savage and James Gately, "Mars Observer Investigation Report Released" (Washington, DC: NASA Headquarters Press Release 94-1, 5 January 1994).

18. Tim Furniss, "Red Light?" *Flight International* (6-12 October 1993): 28-29.

19. Kent Joosten, personal communication.

20. Donald Savage, James Hartsfield, and David Salisbury, "Meteorite Yields Evidence of Primitive Life on Early Mars" (NASA Headquarters Press Release 96-160, 7 August 1996).

21. Everett Gibson, David McKay, Kathie Thomas-Keprta, Christopher Romanek, "The Case for Relic Life on Mars," *Scientific American* (December 1997): 58-65.

22. Kent Joosten, personal communication.

23. Associate Administrators for HEDS Enterprise and Associate Administrator for Space Science Enterprise to Director, Jet Propulsion Laboratory, and Director, Lyndon B. Johnson Space Center, "Integration of Mars Exploration Study and Planning," 7 November 1996, p. 1.

24. *Ibid.*, pp. 1-2.

25. *Ibid.*, p. 2.

26. Douglas Isbell and Michael Braukus, "Space Science and Human Space Flight Enterprises Agree to Joint Robotic Mars Lander Mission" (NASA Headquarters Press Release 97-51, 25 March 1997).

27. Stephen Hoffman and David Kaplan, editors, *Human Exploration of Mars: The Reference Mission of the NASA Mars Exploration Study Team* (Houston: NASA SP-6017, July 1997).

28. *Ibid.*, pp. 1-36 - 1-37, 1-41.

29. *Ibid.*, p. v.

30. Kent Joosten, et al.

31. Michael Duke, editor, *Mars Surface Mission Workshop*, LPI Contribution 934 ( Houston: Lunar and Planetary Institute, 1998).

32. Mars Exploration Study Team, "Mars Exploration Study Program: Report of the Architecture Team" (presentation materials, 6 April 1999), p. 6. The three-pronged approach to Mars exploration apparently dates from a March 1995 NASA Solar System Exploration Subcommittee meeting (Don Bogard, personal communication); it became widely applied to NASA Mars planning only after the McKay team's announcement in August 1996.

33. Duke, *Mars Surface Mission Workshop*, p. 8.

34. Bret Drake, editor, *Reference Mission Version 3.0, Addendum to the Human Exploration of Mars: The Reference Mission of the NASA Mars Exploration Study Team*, EX13-98-036 (Houston: NASA Johnson Space Center Exploration Office, June 1998), pp. 33-37.

35. David S. F. Portree, "Walk This Way," *Air & Space Smithsonian* (October/November 1998): 45-46.

36. Nancy Anne Budden and Michael B. Duke, editors, *HEDS-UP Mars Exploration Forum*, LPI Contribution 955 (Houston: Lunar and Planetary Institute, 1998).

37. Michael Duke, editor, *Second Annual HEDS-UP Forum*, LPI Contribution 979 ( Houston: Lunar and Planetary Institute, 1999).

# Bibliography

## Bibliography

"Ad Astra per Aspera," *Aviation Week & Space Technology* (9 February 1970): 10.

Administrator to Associate Administrator for Manned Space Flight, "Termination of the Contract for Procurement of Long Lead Time Items for Vehicles 516 and 517," 1 August 1968, Logsdon, gen. ed., *Exploring the Unknown*, Vol. I. Washington, DC: NASA SP-4407, 1995, pp. 494-95.

Aeronutronic Division, Ford Motor Company. *EMPIRE, A Study of Early Manned Interplanetary Expeditions*. Huntsville, AL: NASA CR-51709, 21 December 1962.

"Against the Tide," *Aviation Week & Space Technology* (17 March 1969): 15.

Aldrich, Arnold, to Distribution, "Background Material on Cooperation with NPO Energia," 29 June 1992.

Aldrin, Buzz. "The Mars Transit System," *Air & Space Smithsonian*, (October/November 1990): 40-47.

Alter, Dave. "Space Operations Center." NASA Johnson Space Center Press Release 82-008, 19 February 1982.

Ash, R. L., W. L. Dowler, and G. Varsi. "Feasibility of Rocket Propellant Production on Mars," *Acta Astronautica* (July-August 1978): 705-24.

Associate Administrators for HEDS Enterprise and Associate Administrator for Space Science Enterprise to Director, Jet Propulsion Laboratory and Director, Lyndon B. Johnson Space Center, "Integration of Mars Exploration Study and Planning." 7 November 1996.

Baillie, R. F., to R. W. Johnson. "Manned Planetary Exploration Action Item from the Wallops Workshop," 1 August 1978.

"Beggs Calls for Start on Space Station," *Space News Roundup* (25 June 1982): 1, 3-4.

Bekey, Ivan. "A Smaller Scale Manned Mars Evolutionary Program," IAF-89-494, presented at the 40th Congress of the International Astronautical Federation, Malaga, Spain, 7-12 October 1989.

"Bill of Fare," *Aviation Week & Space Technology* (2 February 1970): 11.

Boeing Aerospace Group, *Integrated Manned Interplanetary Spacecraft Concept Definition*, NASA CR-66558, Vol. 1, *Summary*. Seattle, Washington: Boeing, January 1968.

Brooks, Courtney, James Grimwood, and Loyd Swenson, Jr. *Chariots for Apollo, A History of Manned Lunar Spacecraft*. Washington, DC: NASA SP-4205, 1979.

Budden, Nancy Anne, and Michael B. Duke, editors. *HEDS-UP Mars Exploration Forum*, LPI Contribution 955. Houston: Lunar and Planetary Institute, 1998.

Bureau of the Budget, "National Aeronautics and Space Administration: Highlight Summary," 30 October 1968, in Logsdon, gen. ed., *Exploring the Unknown*, Vol. I, pp. 495-99.

"Bush Calls for Two Proposals for Missions to Moon, Mars," *Aviation Week & Space Technology* (12 March 1990): 18-19.

# Bibliography

"Centers Reviewed," *Aviation Week & Space Technology* (19 January 1970): 16.

Chief, NASA Kennedy Space Center Advanced Programs Office, to Distribution, "Minutes of Joint Action Group Meeting of June 29-30, 1966," 8 July 1966.

Clark, Benton. "The Viking Results—The Case for Man on Mars," AAS 78-156, in Richard Johnston, Albert Naumann, and Clay Fulcher, editors, *The Future U.S. Space Program*. San Diego: Univelt, Inc., pp. 263-78.

Collins, Michael. *Mission to Mars*. New York: Grove Weidenfeld, 1990.

Committee on the Human Exploration of Space, *Human Exploration of Space: A Review of NASA's 90-Day Study and Alternatives*. Washington, DC: National Academy Press, 1990.

"Conferees Vote Space Cut," *Aviation Week & Space Technology* (7 August 1967): 24.

"Congressional Critics Aim to Cut NASA Budget to $4-Billion Level," *Aviation Week & Space Technology* (12 February 1968): 22-23.

Contracting Officer to Prospective Contractors, "Planetary Surface Sample Return Probe Study for Manned Mars/Venus Reconnaissance/Retrieval Missions." Request for Proposal No. BG721-28-7-528P, 3 August 1967.

"Cost of Tet," *Aviation Week & Space Technology* (27 May 1968): 25.

Covault, Craig. "Exploration Initiative Work Quickens as Some Concepts Avoid Station," *Aviation Week & Space Technology* (17 September 1990): 36-37.

Covault, Craig. "Launch Capacity, EVA Concerns Force Space Station Re-Design," *Aviation Week & Space Technology* (21 July 1986): 18-20.

Covault, Craig. "Ride Panel Calls for Aggressive Action to Assert U.S. Leadership in Space," *Aviation Week & Space Technology* (24 August 1987): 26-27.

Covault, Craig. "Ride Panel Will Urge Lunar Base, Earth Science as New Space Goals," *Aviation Week & Space Technology* (13 July 1987): 16-18.

Covault, Craig. "Space Policy Outlines Program to Regain U.S. Leadership," *Aviation Week & Space Technology* (22 February 1988): 20-21.

Covault, Craig. "White House Endorses Plan for Shuttle, Station Scale-Back," *Aviation Week & Space Technology* (17 December 1990): 18-20.

Covington, Clarke. "The Role of the Space Operations Center," presentation materials, 28 May 1981.

Cravotta, Charles, and Melanie DeForth. "Soviet Plans for a Manned Flight to Mars." Office of Scientific and Weapons Research, U.S. Central Intelligence Agency, 2 April 1985.

Crocco, Gaetano A. "One-Year Exploration-Trip Earth-Mars-Venus-Earth," Rendiconti del VII Congresso Internanzionale Astronautico, Associazione Italiana Razzi, 1956, presented at the Seventh Congress of the International Astronautical Federation, Rome, Italy, 1956, pp. 227-52.

Crossley, Louise. *Explore Antarctica*. Cambridge, England: Cambridge University Press, 1995.

"Darman Backs NASA," *Aviation Week & Space Technology* (21 May 1990): 17.

David, Leonard. "Faster, Cheaper Mars Exploration Proposed," *Space News* (11-17 June 1990): 4, 37.

Day, Dwayne. "Doomed to Fail," *Spaceflight* (March 1995): 79-83.

Day, Dwayne. "Viewpoint: Paradigm Lost," *Space Policy* (August 1995): 153-59.

Dewar, James. "Atomic Energy: The Rosetta Stone of Space Flight," *Journal of the British Interplanetary Society* (May 1994): 199-206.

Dixon, Franklin. "Manned Planetary Mission Studies from 1962 to 1968," IAA-89-729, presented at the 40th Congress of the International Astronautical Federation, Malaga, Spain, 7-12 October 1989.

Dixon, Franklin. "Summary Presentation: Study of a Manned Mars Excursion Module," *Proceeding of the Symposium on Manned Planetary Missions: 1963/1964 Status*, Huntsville, AL, 1964, pp. 443-523.

Drake, Bret, editor. *Reference Mission Version 3.0, Addendum to the Human Exploration of Mars: The Reference Mission of the NASA Mars Exploration Study Team*, EX13-98-036. Houston: NASA Johnson Space Center Exploration Office, June 1998.

Duke, Michael, and Nancy Anne Budden, editors. *Mars Exploration Study Workshop II*. Houston: NASA CP-3243, November 1993.

Duke, Michael, and Paul Keaton, editors. *Manned Mars Missions, Working Group Summary Report*. Huntsville, AL, and Los Alamos, NM: NASA M001, NASA/LANL, May 1986.

Duke, Michael, and Paul Keaton, editors. *Manned Mars Missions, Working Group Papers*, Vol. 1 and Vol. 2. Huntsville, AL, and Los Alamos, NM: NASA M002, NASA/LANL, June 1986.

Duke, Michael, editor. *Mars Surface Mission Workshop*. LPI Contribution No. 934, Houston: Lunar and Planetary Institute, 1998.

Duke, Michael, editor. *Second Annual HEDS-UP Forum*, LPI Contribution 979, Houston: Lunar and Planetary Institute, 1999.

Duke, Michael, Wendell Mendell, and Barney Roberts. "Lunar Base: A Stepping Stone to Mars," AAS 84-162, Christopher McKay, editor, *The Case for Mars II*. Houston: Univelt, Inc., 1985, pp. 207-220.

Ezell, Edward. "Man on Mars: The Mission That NASA Did Not Fly," presented at the American Association for the Advancement of Science Annual Meeting, Houston, TX: 3-8 January 1979.

Ezell, Edward, and Linda Neumann Ezell. *On Mars: Exploration of the Red Planet 1958-1978*. NASA, 1984.

Faget, Maxime, and Paul Purser. "From Mercury to Mars," *Aeronautics & Aerospace Engineering* (February 1963): 24-28.

# Bibliography

Farquhar, Robert, and David Dunham. "Libration-Point Staging Concepts for Earth-Mars Transportation," *Manned Mars Missions: Working Group Papers*, Vol. 1. Huntsville, AL, and Los Alamos, NM: NASA M002, NASA/LANL, June 1986, pp. 66-77.

Friedman, Louis. "Visions of 2010," *The Planetary Report* (March/April 1985): 4-6, 22.

Furniss, Tim. "Red Light?" *Flight International* (6-12 October 1993): 28-29.

Furniss, Tim. *Space Shuttle Log*. London: Jane's, 1986.

"Future Efforts to Stress Apollo Hardware," *Aviation Week & Space Technology* (16 November 1964): 48-51.

General Dynamics Astronautics. *A Study of Early Manned Interplanetary Missions Final Summary Report*. General Dynamics Astronautics, 31 January 1963.

Gibson, Everett, David McKay, Kathie Thomas-Keprta, Christopher Romanek. "The Case for Relic Life on Mars ," *Scientific American* (December 1997): 58-65.

Glasstone, Samuel. *The Book of Mars*. Washington, DC: NASA SP-179, 1968.

Glushko, V., Y. Semyonov, and L. Gorshkov. "The Way to Mars," *The Planetary Report* (November December 1988): 4-8.

Gray, Edward, and Franklin Dixon. "Manned Expeditions to Mars and Venus," Eric Burgess, editor, *Voyage to the Planets*. San Diego: Univelt, Inc., 1967, pp. 107-35.

Gray, Edward, to H. K. Weidner, F. L. Williams, J. W. Carter, R. J. Harris, J. P. Claybourne, R. Hock, and R. J. Cerrato, TWX. "Follow-on Activity for Manned Planetary Program." 2 December 1966.

Gray, Edward, to H. K. Weidner, F. L. Williams, M. Faget, W. E. Stoney, J. West, J. P. Claybourne, and R. Hock, TWX. "Meeting to Establish Follow-on Activities Covering the Advanced Manned Planetary, Earth Orbital, and Lunar Exploration Programs." 17 November 1966.

Guilmartin, Patricia. "House Kills Funding for Moon/Mars Effort, " *Aviation Week & Space Technology* (2 July, 1990): 28.

Hammock, David, and Bruce Jackson. "Vehicle Design for Mars Landing and Return to Mars Orbit, " in George Morgenthaler, editor, *Exploration of Mars*. San Diego: Univelt, Inc., 1964, pp. 174-92.

Hansen, James. *Enchanted Rendezvous: John C. Houbolt and the Genesis of the Lunar-Orbit Rendezvous Concept*. Washington, DC: NASA Monographs in Aerospace History #4, 1995.

Hartmann, William, and Odell Raper. *The New Mars: The Discoveries of Mariner 9*. Washington, DC: NASA SP-337, 1974.

Henderson, Breck. "Livermore Plan for Exploring Moon, Mars Draws Space Council Attention," *Aviation Week & Space Technology* (22 January 1990): 84-85, 88.

Heppenheimer, T. A. *The Space Shuttle Decision: NASA's Search for a Reusable Space Vehicle.* Washington, DC: NASA SP-4221, 1999.

Hess, Mark. "NASA Proceeding Toward Space Station Development." NASA Johnson Space Center Press Release 87-50, April 3, 1987.

Hill, Arthur. "Apollo Shape Dominates NAR Manned Mars Study," *Aerospace Technology* (6 May 1968):. 26-27.

Himmel, S. C., J. F. Dugan, R. W. Luidens, and R. J. Weber. "A Study of Manned Nuclear-Rocket Missions to Mars," IAS Paper No. 61-49, presented at the 29th Annual Meeting of the Institute of Aerospace Sciences, 23-25 January 1961.

Hoffman, Stephen and David Kaplan, editors. *Human Exploration of Mars: The Reference Mission of the NASA Mars Exploration Study Team.* Houston: NASA SP-6017, July 1997.

Hotz, Robert. "New Era for NASA," *Aviation Week & Space Technology* (7 August 1967): 17.

Hotz, Robert. "The Endless Frontier," *Aviation Week & Space Technology* (11 August 1969): 17.

"House Unit Trims NASA Budget, Fight Pledged for Further Slashes," *Aviation Week & Space Technology* (22 May 1967): 24.

House, William. "The Development of Nuc lear Rocket Propulsion in the United States ," *Journal of the British Interplanetary Society*, Vol. 19, No. 8 (March-April, 1964): 306-318.

Hyde, R. A., M. Y. Ishikawa, and L. L. Wood. "Toward a Permanent Lunar Settlement in the Coming Decade: The Columbus Project," Livermore, CA: UCRL-93621, DE86 006709, Lawrence Livermore National Laboratory , 19 November 1985.

Hyde, Rod, Yuki Ishikawa, and Lowell Wood. "An American-Traditional Space Exploration Program: Quick, Inexpensive, Daring, and Tenacious, Briefing to the National Space Council." Livermore, CA: LLNL Doc. No. Phys. Brief 89-403, September 1989.

Hyde, Roderick, Muriel Ishikawa, and Lowell Wood. "Mars in this Century: The Olympia Project," UCRL-98567, DE90 008356, Lawrence Livermore National Laboratory, presented at the U. S. Space Foundation 4th National Space Symposium, Colorado Springs, Colorado, 12-15 April 1988.

Isbell, Douglas. "Congress Says OK to Moon, Mars Work," *Space News* (28 May-3 June 1990): 3, 20.

Isbell, Douglas. "Ex-Astronaut Stafford to Head Moon-Mars Outreac h Team," *Space News* (4-10 June 1990): 1.

Isbell, Douglas, and Andrew Lawler. "Senators Assail Bush Plan," *Space News* (7-13 May 1990): 1, 27.

Isbell, Douglas, and Michael Braukus. "Space Science and Human Space Flight Enterprises Agree to Joint Robotic Mars Lander Mission." NASA Headquarters Press Release 97-51, 25 March 1997.

Jenkins, Morris. *Manned Exploration Requirements and Considerations.* Houston: NASA Manned Spacecraft Center, February 1971.

# Bibliography

Johnsen, Katherine. "NASA Gears for $4-Billion Fund Limit," *Aviation Week & Space Technology* (27 May 1968): 30-31.

Johnsen, Katherine. "Webb Refuses to Choose Program for Cuts," *Aviation Week & Space Technology* (31 July 1967): 20.

Joosten, Kent, Ryan Schaefer, and Stephen Hoffman. "Recent Evolution of the Mars Reference Mission," AAS-97-617, presented at the AAS/AIAA Astrodynamic Specialist Conference, Sun Valley, Idaho, 4-7 August 1997.

Keaton, Paul. *A Moon Base / Mars Base Transportation Depot*. Los Alamos, NM: LA-10552-MS, UC-34B, Los Alamos National Laboratory, September 1985.

Klass, Philip, "Commission Considers Joint Mars Exploration, Lunar Base Options," *Aviation Week & Space Technology* (29 July 1985): 47-48.

Launius, Roger. "The Waning of the Technocratic Faith: NASA and the Politics of the Space Shuttle Decision," in Philippe Jung, editor, *History of Rocketry and Astronautics*, AAS History Series, Volume 21. San Diego, CA: Univelt, Inc., 1997, pp. 179-96.

Lawler, Andrew. "Bush: To Mars by 2019," *Space News* (14-20 May 1990): 1, 20.

Lawler, Andrew. "Bush Moon-Mars Plan Handed First Defeat," *Space News* (18-24 June 1990): 3, 21.

Lewis, Richard. "On the Golden Plains of Mars," *Spaceflight* (October 1976): 356-364.

Lewis, Richard. "The Puzzle of Martian Soil," *Spaceflight* (November 1976): 391-395.

Ley, Willy, and Wernher von Braun. *The Exploration of Mars*. New York: Viking Press, 1956.

Lockheed Missiles & Space Company. *Manned Interplanetary Mission Study*. Sunnyvale, CA: Lockheed Missiles and Space Company, March 1963.

Logsdon, John. *The Decision to Go to the Moon: Project Apollo and the National Interest*. Boston: MIT Press, 1970.

Logsdon, John. "From Apollo to the Space Shuttle: U.S. Space Policy, 1969-1972," unpublished manuscript.

Logsdon, John, gen. ed., with Linda Lear, Janelle Warren-Findlay, Ray Williamson, and Dwayne Day. *Exploring the Unknown: Selected Documents in the History of the U .S. Civil Space Program, Volume 1: Organizing for Exploration*. Washington, DC: NASA SP-4407, 1995.

Mandell, Humboldt. "Space Station - The First Step," AAS 84-160, in Christopher McKay, editor, *The Case for Mars II*. San Diego: Univelt, Inc., 1985, pp. 157-70.

Mann, Paul. "Commission Sets Goals for Moon, Mars Settlement in 21st Century ," *Aviation Week & Space Technology* (24 March 1986) 18-21.

Mars Exploration Study Team. "Mars Exploration Study Program: Report of the Architecture Team," presentation materials, 6 April 6, 1999.

Martin Marietta. *Manned Mars System Study (MMSS) Executive Summary*. Denver: Martin Marietta, July 1990.

Martin Marietta. *Manned Mars System Study (Mars Transportation and Facility Infrastructure Study), Volume II, Final Report*. Denver: Martin Marietta, July 1990.

Mayo, Robert, Director, Bureau of the Budget. "Memorandum for the President, 'Space Task Group Report,'" 25 September 1969, Logsdon, gen. ed., *Exploring the Unknown*, Vol. 1, pp. 544-46.

Merrifield, Robert. "A Historical Note on the Genesis of Manned Interplanetary Flight," AAS Preprint 69-501, presented at the AAS 15th Annual Meeting, 17-20 June 1969.

Meyerson, Harvey. "Spark Matsunaga 1916-1990," *The Planetary Report* (July/August 1990): 26.

Mueller, George, to John Naugle, October 6, 1969.

"NASA Accelerates Lunar Base Planning as Station Changes Draw European Fire," *Aviation Week & Space Technology* (18 September 1989): 26-27.

NASA. *America's Next Decades in Space: A Report to the Space Task Group*. Washington, DC: NASA September 1969.

NASA. *Astronautics and Aeronautics 1966*. NASA SP-4007.

NASA. *Astronautics and Aeronautics 1968*. NASA SP-4010.

NASA. *Astronautics and Aeronautics 1969*. NASA SP-4014.

NASA. *Astronautics and Aeronautics 1970*. NASA SP-4015.

NASA. *Astronautics and Aeronautics 1972*. NASA SP-4017.

NASA. *Astronautics and Aeronautics 1986-1990*. Washington, DC: NASA SP-4027.

"NASA Budget Faces House-Senate Parley," *Aviation Week & Space Technology* (29 September 1969): 19.

"NASA Forms Office to Study Manned Lunar Base, Mars Missions," *Aviation Week & Space Technology* (8 June 1987): 22.

"NASA Funds $100-Million Pathfinder Program for Mars, Lunar Technology," *Aviation Week & Space Technology* (18 January 1988): 17.

NASA. "Integrated Manned Space Flight Program, 1970-1980." Washington, DC: NASA, 12 May 1969.

"NASA Managers Divided on Station," *Aviation Week & Space Technology* (28 July 1986) 24-25.

NASA. "NASA Facts: Space Station." Kennedy Space Center Press Release No. 16-86, January 1986.

NASA. "Outline of NASA Presentation to Space Task Group, August 4, 1969." 28 July 1969.

# Bibliography

NASA. "Part 17: Panel Discussion," *Proceeding of the Symposium on Manned Planetary Missions:   1963/1964 Status.* Huntsville, AL: NASA, 1964, pp. 737-54.

NASA. *Proceedings of the Seventh Annual Working Group on Extraterrestrial Resources* . Washington, DC: NASA SP-229, 1970.

NASA. *Space Station Freedom Media Handbook.* Washington, DC: NASA, April 1989.

NASA. "Statement by Dr. Sally K. Ride, Associate Administrator for Exploration (Acting) before the Subcommittee on Space Science and Applications, Committee on Science, Space, and Technology, House of Representatives," 22 July 1987.

NASA. "A Report from Mariner IV," *NASA Facts* 3, No. 3 (1966).

NASA. *Report of the 90-Day Study on Human Exploration of the Moon and Mars.* "Cost Summary," unpublished chapter.

NASA. *Report of the 90-Day Study on Human Exploration of the Moon and Mar*      s. Washington, DC: NASA, November 1989.

National Commission on Space (NCOS). *Pioneering the Space Frontier: The Report of the National Commission on Space.* New York: Bantam Books, May 1986.

Nicks, Oran. *Summary of Mariner 4 Results* . Washington, DC: NASA SP-130.

Nixon, Richard. "Memorandum for the Vice President, the Secretary of Defense, the Acting Administrator, NASA, and the Science Advisor," 13 February 1969, Logsdon, gen. ed., *Exploring the Unknown*, Vol. 1, pp. 512-513.

Normyle, William. "Manned Mars Flights Studied for the 1970s ," *Aviation Week & Space Technology* (27 March 1967): 62-63, 65, 67.

Normyle, William. "Manned Mission to Mars Opposed," *Aviation Week & Space Technology* (18 August 1969): 16-17.

Normyle, William. "NASA Aims at 100-man Station," *Aviation Week & Space Technology* (24 February 1969): 16-17.

Normyle, William. "NASA Plans Five-Year Fund Rise," *Aviation Week & Space Technology* (14 October 1968): 16-17.

Normyle, William. "Post-Apollo Program Potential Emerging," *Aviation Week & Space Technology* (6 March 1967): 126-131.

Normyle, William. "Priority Shift Blocks Space Plans," *Aviation Week & Space Technology* (11 September 1967): 26-27.

Normyle, William. "Small Hope Seen to Restore Space Funds ," *Aviation Week & Space Technology* (10 July 1967): 38-39.

North American Rockwell Corporation Space Division. *Definition of Experimental  Tests for a Manned  Mars Excursion Module: Final Report, Volume I, Summary.* SD 67-755-1, 12 January 1968.

"Notice to NASA," *Aviation Week & Space Technology* (15 January 1990): 15.

Oberg, Alcestis. "The Grass Roots of the Mars Conference," AAS 81-225, Penelope Boston, editor, *The Case for Mars*, San Diego: Univelt, Inc., 1984, pp. ix-xii.

Office of Exploration. *Exploration Studies Technical Report, FY 1988 Status, Volume 1: Technical Summary.* Washington, DC: NASA TM-4075, December 1988.

Office of Exploration. "FY88 Exploration Studies Technical Presentation to the Administrator," presentation materials, 25 July 1988.

Office of the White House Press Secretary. "Remarks of the President at the 20th Anniversary of Apollo Moon Landing." 20 July 1989.

O'Lone, Richard. "Scientist Sees Space Station Useful Only if Linked to Manned Mars Mission," *Aviation Week & Space Technology* (25 January 1988): 55, 57.

"OMB Limits NASA to $15 million for NERVA," *Aviation Week & Space Technology* (4 October 1971): 20.

Ordway, Frederick and Mitchell Sharpe. *The Rocket Team*. New York: Thomas Y. Crowell, 1979.

Paine, Thomas. "Overview: Report of the National Commission on Space," in Duke Reiber, *The NASA Mars Conference*. San Diego: Univelt, Inc., 1988, pp. 525-34.

Paine, Thomas. "Problems and Opportunities in Manned Space Flight," Logsdon, gen. ed., *Exploring the Unknown*, Vol. 1, p. 513-19.

Paine, Thomas. "A Timeline for Martian Pioneers," AAS 84-150, in Christopher McKay, editor, *The Case for Mars II*. San Diego: Univelt, Inc., 1985, pp. 3-21.

Paine, Thomas. "Who Will Lead the World's Next Age of Discovery?" *Aviation Week & Space Technology* (21 September 1987): 43-45.

Planetary JAG. *Planetary Exploration Utilizing a Manned Flight System.* Washington, DC: NASA, 1966.

Portree, David S. F. *NASA's Origins and the Dawn of the Space Age.* Washington, DC: NASA Monographs in Aerospace History #10, 1998.

Portree, David S. F. *Thirty Years Together: A Chronology of U.S.-Soviet Space Cooperation.* Houston: NASA CR-185707, February 1993.

Portree, David S. F. "Walk this Way," *Air & Space Smithsonian* (October/November 1998): 44-47.

"President Bush Sets 2019 Manned Mars Objective," *Aviation Week & Space Technology* (21 May 1990):19.

*Presidential Papers of the President: Administration of Ronald Reagan, 1985.* Washington, DC: U.S. Government Printing Office, 1985.

President's Science Advisory Committee. *The Space Program in the Post-Apollo Period.* Washington, DC: The White House, February 1967.

# Bibliography

Ragsdale, Lyn. "Politics Not Science: The U.S. Space Program in the Reagan and Bush Years," *Spaceflight and the Myth of Presidential Leader ship*, Roger Launius and How ard McCurdy, editors. Urbana: University of Illinois Press, 1997, pp. 133-71.

Rall, Charles and Walter Hollister. "Free-fall Periodic Orbits Connecting Earth and Mars ," AIAA No. 71-92, presented at the American Institute of Aeronautics and Astronautics 9th Aerospace Sciences Meeting, New York, New York, January 25-27, 1971.

"Reaching Out," *Aviation Week & Space Technology* (4 June 1990): 15.

Ride, Sally. *Leadership and America's Future in Space*. Washington, DC: NASA, August 1987.

Roberts, Barney. "Concept for a Manned Mars Flyby ," *Manned Mars Missions: Working Group Papers*, Vol.1. Huntsville, AL, and Los Alamos, NM: NASA M002, NASA/LANL, June 1986, pp. 203-18.

Ruppe, Harry. *Manned Planetary Reconnaissance Mission Study: Venus / Mars Flyby*. Huntsville, AL: NASA TM X-53205, 1965.

Sagan, Carl. "To Mars," *Aviation Week & Space Technology* (8 December 1986) 10.

Savage, Donald, and James Gately. "Mars Observer Investigation Report Released. " NASA Headquarters Press Release 94-1, January 5, 1994.

Savage, Donald, James Hartsfield, and David Salisbury. "Meteorite Yields Evidence of Primitive Life on Early Mars." Washington, DC: NASA Headquarters Press Release 96-160, 7 August 1996.

Schlesinger, Jr., Arthur. *The Almanac of American History*. Greenwich, CT: Brompton Books, 1993.

Schmitt, Harrison. "A Millennium Project—Mars 2000, " in Wendell Mendell, editor, *Lunar Bases and Space Activities of the 21st Century*. Houston: Lunar and Planetary Science Institute, 1985, pp. 787-93.

"Science Advisors Urge Balanced Program," *Aviation Week & Space Technology* (6 March 1967): 133, 135, 137.

Science Applications International Corporation. *Piloted Sprint Missions to Mar s*. Schaumberg, IL: Report No. SAIC-87/1908, Study No. 1-120-449-M26, November 1987.

Scientific Industrial Corporation "Energia." *Mars Manned Mission: Scientific / Technical Report*. Moscow: USSR Ministry of General Machinery, 1991.

"Scientists Urge Priority for Mars Missions," *Aviation Week & Space Technology* (23 November 1964): 26-27.

Seamans, Robert, Secretary of the Air Force, to Spiro Agnew, Vice President, letter, August 4, 1969, Logsdon, gen. ed., *Exploring the Unknown*, Vol. 1, pp. 519-22.

SEI Synthesis Group . *America at the Threshold: America's Space Exploration Initiative* . Washington, DC: Government Printing Office, May 1991.

Semyonov, Yuri, and Leonid Gorshkov. "Destination Mars," *Science in the USSR* (July-August 1990): 15-18.

"Senior Soviet Space Officials Outline Plan for J oint Mars Mission," *Aviation Week & Space Technology* (19 November 1990): 67.

Shifrin, Carole. "NASA Nears Final Decisions on Station Configuration," *Aviation Week & Space Technology* (10 March 1986): 107-09.

Singer, S. Fred. "To Mars By Way of Its Moons," *Scientific American* (March 2000): 56-57.

Singer, S. Fred. "The PH-D Proposal: A Manned Mission to Phobos and Deimos," AAS 81-231, in Penelope Boston, editor, *The Case for Mars*. San Diego: Univelt, Inc., 1984, pp. 39-65.

Smith, J. N. *Manned Mars Missions in the Unfavorable (1975-1985) Time Period: Executive Summary Report*. Huntsville, AL: NASA TM X-53140, 1964.

Sohn, Robert. "A Chance for an Early Manned Mars Mission," *Astronautics & Aeronautics* (May 1965): 28-33.

Sohn, Robert. "Summary of Manned Mars Mission Study," *Proceeding of the Symposium on Manned Planetary Missions: 1963/1964 Status*. Huntsville, AL: NASA TM X-53049, 1964. pp. 149-219.

"Space Funds Cut Deeply by House, Senate," *Aviation Week & Space Technology* (3 July 1967): 28.

"Space in the 1970s," *Aviation Week & Space Technology* (9 February 1970): 11.

"Space Manpower," *Aviation Week & Space Technology* (11 August 1969): 25.

"Space Policy," *Aviation Week & Space Technology* (30 October 1989): 15.

Space Science and Technology Panel of the President's Science Advisory Committee. *The Next Decade in Space*. Washington, DC: Executive Office of the President, Office of Science and Technology, March 1970.

Space Sciences Department. *Manned Lunar, Asteroid and Mars Missions, Visions of Space Flight: Circa 2001*. Shaumberg, IL: Science Applications International Corporation, September 1984.

Space Task Group. *The Post-Apollo Space Program: Directions for the Future*. Washington, DC: NASA, September 1969.

"Space Wraith," *Aviation Week & Space Technology* (24 July 1989): 21.

Spacecraft Engineering Branch. *Apollo-based Venus/Mars Flybys*. Houston: NASA MSC, September 1967.

"Spaced Out," *Aviation Week & Space Technology* (15 September 1986): 11.

Spitzer, Cary, editor. *Viking Orbiter Views of Mars*. Washington, DC: NASA, 1980.

Stockton, William, and John Noble Wilford. *Spaceliner*. New York: Times Books, 1981.

Stone, Irving. "Manned Planetary Vehicle Study Proposed," *Aviation Week & Space Technology* (2 October 1967): 87, 90.

# Bibliography

Stuhlinger, Ernst. "Possibilities of Electrical Space Ship Propulsion," in Friedrich Hecht, editor, *Bericht über de V Internationalen Astronautischen Kongress,* Osterreichen Gesellschaft für Weltraumforschung, 1955, pp. 100-19.

Stuhlinger, Ernst, and Joseph King. "Concept for a Manned Mars Expedition with Electrically Propelled Vehicles," *Progress in Astronautics*, Vol. 9. San Diego: Univelt, Inc., 1963, pp. 647-663.

Taylor, Hal, "LBJ Wants Post-Apollo Plans," *Missiles and Rockets* (4 May 1964): 12.

"The Mars Declaration," special supplement to *The Planetary Report*, November/December 1987.

Titus, R. R., "FLEM—Flyby-Landing Excursion Mode ," AIAA Paper No. 66-36, presented at the 3rd AIAA Aerospace Sciences Meeting, New York, New York, 24-26 January 1966.

Townes, Charles, et al. "Report of the Task Force on Space," 8 January 1969, Logsdon, gen. ed., *Exploring the Unknown*, Vol. 1, pp. 499-512.

Truly, Richard and Franklin Martin. "Briefing to NASA Employees." 26 July 1989.

University of Texas and Texas A&M University Design Team. "To Mars—A Manned Mars Mission Study," Summer Project Report, NASA Universities Advanced Space Design Program, Advanced Programs Office. Houston: Johnson Space Center, August 1985.

"U.S. Astronaut to Visit Soviet Station, Cosmonaut to Fly on Shuttle ," *Aviation Week & Space Technology* (22 October 1990): 24-25.

U.S. Congress. *Nuclear Rocket Development Program,* Joint Hearings before the Committee on Aeronautical and Space Sciences, United States Senate and the J oint Committee on Atomic Energy, 92nd Congress of the United States, First Session, 23-24 February 1971.

"U.S. Space Funding to Grow Moderately," *Aviation Week & Space Technology* (6 March 1967): 123-26.

von Braun, Wernher. "Crossing the Last Frontier," *Collier's* (22 March 1952). 24-31.

von Braun, Wernher. "Man on the Moon: The Journey," *Collier's* (18 October 1952): 52-60.

von Braun, Wernher. "Manned Mars Landing Presentation to the Space Task Group," presentation materials, 4 August 1969.

von Braun, Wernher. *The Mars Project*. Urbana, IL: University of Illinois Press, 1962.

von Braun, Wernher. "The Next 20 Years of Interplanetary Exploration," *Astronautics & Aeronautics* (November 1965): 24-34.

von Braun, Wernher, and Cornelius Ryan. "Can We Get to Mars?" *Collier's* (30 April 1954): 22-29.

"Washington Roundup," *Aviation Week & Space Technology* (21 July 1969): 15.

Watts, Raymond. "Manned Exploration of Mars?" *Sky & Telescope* (August 1963): 63-67, 84.

Weaver, David and Mic hael Duke. "Mars Exploration Strategies: A Reference Program and Comparison of Alternative Architectures," AIAA 93-4212, presented at the AIAA Space Program and Technologies Conference, Huntsville, Alabama, 21-23 September 1993.

"Webb Urges Full $4-Billion NASA Fund," *Aviation Week & Space Technology* (1 July 1968): 22.

Welch, S. M. and C. R. Stoker, editors. *The Case for Mars: Concept Development for a Mar s Research Station.* Boulder, CO: Boulder Center for Science Policy, 10 April 1986.

Whipple, Fred, and Wernher von Braun. "Man on the Moon: The Exploration," *Collier's* (18 October 1952): 38-48.

"White House Stand Blocks NASA Budget Restoration," *Aviation Week & Space Technology* (28 August 1967): 32.

Wilks, Willard and Rex Pay. "Quest for Martian Life Re-Emphasized," *Technology Week* (6 June 1966): 26-28.

Wilson, Andrew. *Solar System Log*. London: Jane's, 1987.

"Work on Future Saturn Launchers Halted," *Aviation Week & Space Technology* (12 August 1968): 30.

Zubrin, Robert, and David Baker. "Humans to Mars in 1999," *Aerospace America* (August 1990): 30-32, 41.

Zubrin, Robert, and David Weaver. "Practical Methods for Near-Term Piloted Mars Missions," AIAA 93-2089, presented at the AIAA/SAE/ASME/ASEE 29th Joint Propulsion Conference, Monterey, California, 28-30 June 1993.

Zubrin, Robert, with Richard Wagner. *The Case for Mars*. New York: Free Press, 1996.

# About the Author

David S. F. Portree is an Arizona-based science writer and historian. His other NASA history publications include Orbital Debris: A Chronology (with Joseph P. Loftus, Jr., 1999), NASA's Origins and the Dawn of the Space Age (1998), Walking to Olympus: An EVA Chronology (with Robert C. Trevino, 1997), and Mir Hardware Heritage (1995).

# Index

# Index

# Index

Kraft, Christopher, 85

Lagrange, Joseph, 64
Lagrange point station, 63, 68; as stepping stone to Mars, 64
Laird, Melvin, 42
Langley Research Center, 8, 15, 20, 32, 36, 42; supports ISRU research, 56
Launius, Roger, 41
*Leadership and America's Future in Space* (Ride Report), 69-73, 81, 89
Lewis Research Center, 5, 6, 8, 9, 15, 19, 34, 85, 97, 98; and first NASA Mars study, 5-6, 37; and Design Reference Mission, 97-98; *see also* NASA—Glenn Research Center at Lewis Field
Ley, Willy, 2, 3
Life Systems, 73
lifting body, 13, 15, 16, 17, 20, 23
lithium, 85
Lockheed Missiles and Space Company, 12, 13, 14, 29, 35; *see also* EMPIRE
Logsdon, John, vii, 48
Los Alamos National Laboratory (LANL)— *see under* Atomic Energy Commission (AEC)
*Los Angeles Herald-Examiner*, 43
Lovell, James, 41
Low, George, 41, 49, 52, 74
Lowell, Percival, 23, 53
Lunar and Planetary Institute, 97
Lunar Bases and Space Activities of the 21st Century, 61
Lunar Orbiter, 53
Lunokhod 2, 95

Manarov, Musa, 74
Mandell, Humboldt, 63, 79
Manned Mars Mission (MMM): study, 51, 61, 84; work shop, 61, 64
Manned Spacecraft Center (MSC), 15, 16, 19, 25, 31, 32, 37, 41, 49, 50, 59, 60, 80; and Planetary Missions Requirements Group, 49-51; *see also* Johnson Space Center
Margaritifer Sinus region, 3
Mariner, 26, 53; Mariner 2, 12, 23; Mariner 4, 22, 23, 24, 25, 30, 37, 44; Mariner 6, 43-44, 53; Mariner 7, 43-44, 46, 53; Mariner 9, 53, 54, 55, 95
Mars: affect of environment on unprotected human, 54; atmosphere, 16, 17, 20, 21, 23, 24, 25, 37, 38, 54-56, 67, 86, 90; canals, 2-4, 23; channels, 53, 54, 95; dust, 53, 54, 97, 99; opposition, 3, 4, 18, 19, 53, 75; permafrost, 53, 54; poles, 1, 17, 18, 43, 67; popular

image, 23, 53, 54, 55; water, 17, 23, 53, 54, 55, 56, 81, 94, 98; as an abode of life, 2, 3, 15, 17, 23, 26, 29, 45, 51, 53, 54, 71, 89, 94, 97, 98; as base/set tlement site, 2, 19, 21, 45, 55, 56, 60, 62, 63, 70, 71, 79, 80, 81, 89, 90, 91, 92, 93, 97; as revealed by Mariner 4, 23; as revealed by Mariner 9, 53, 55; as revealed by Viking, 54-55
*Mars and Beyond*, 7
*Mars Declaration, The*—*see under* Planetary Society, The
Mars Direct, 85, 89-92; small Earth-Return Vehicle, 91; *see also* Design Reference Mission
Mars Exploration Study Team—*see under* Design Reference Mission
Mars Global Surveyor—*see under* Mars Surveyor Program
Mars Observer, 93, 94, 96
Mars Orbit Rendezvous (MOR), 9, 14, 15, 16, 29, 37, 63, 93; *see also* piloted Mars landers
Mars Pathfinder, 89, 95; renamed Sagan Memorial Station, 95
Mars Surface Sample Return (MSSR) lander— *see under* piloted Mars flyby
Mars Surveyor Program, 94, 95, 96; Mars Global Surveyor, 96, 98
Mars Transportation and Facility Infrastructure Study —*see under* Martin Marietta Corporation
"Mars Underground"—*see under* Case for Mars, The
Marshall Space Flight Center, 7, 9, 11, 12, 15, 18, 20, 21, 24, 25, 35, 44, 49, 55, 61, 73, 89; Future Projects Office, 11, 12, 18, 20, 21, 24, 44; Symposium on Manned Planetary Missions, 16, 21
Martian meteorite—*see* ALH 84001
Martin, Franklin, 77, 78
Martin Marietta Corporation, 15, 56, 73, 80, 85, 89, 90; Mars Transportation and Facility Infrastructure Study, 73, 80
Matsunaga, Spark, 73
Mayo, Robert, 42, 44, 46-48
McGill University, 94
McKay, Christopher, 57, 58, 63
McKay, David, 94
Mendell, Wendell, 63
Mercury (planet), 53
Mercury, 12, 15, 31, 33; *Freedom 7*, 6
Meteor Crater, 98
meteoroids, 8, 13, 14, 17, 19, 23, 62, 97
methane, 38, 39, 55, 56, 90, 91, 92, 93
Mie crater, 54
Miller, George, 47

# Index

# Index

Energy; NASA; United States Air Force; United States Army; White House
University of Arizona, 67
University of Illinois Press, 1
University of Texas, 71
urban riots, 29, 31, 32
Utopia Planitia region, 54

V-2 missile, 1, 7
Valles Marineris, 53
Van Allen Radiation Belts, 6, 8, 58, 97, 98
Varsi, Giulio, 55, 56, 90
Vastitas Borealis region, 17
Venus, 12, 13, 20, 23, 26, 32, 37, 45, 53, 58, 62, 80
Vietnam, 24, 29-31, 39, 49; Tet offensive, 39
Viking, 32, 35, 48, 53-57, 60, 68, 74, 93, 95
von Braun, Wernher, 1-4, 6, 7, 11, 13, 19, 21, 23, 33, 42-48, 98; career apogee, 44; *Das Marsprojekt*, 1, 3; *The Exploration of Mars*, 3; *The Mars Project*, 1, 2, 7, 11, 19
Vostok 1, 6
Voyager Mars/Venus program, 24-26, 29-32, 35; as victim of piloted flyby planning, 32

Wallops Island, 60
*Washington Evening Star*, 41

*Washington Post*, 74
Webb, James, 24, 31, 39, 40, 41
weightlessness, 1, 3, 13, 28, 58, 64, 71, 84, 90
Whipple, Fred, 2
White, Ed, 30
White House, 29, 32, 35, 39, 40, 47, 49, 52, 60, 67, 68, 69, 73, 77, 78, 81, 83; Budget Bureau, 21, 24, 29, 35, 40, 42, 48; National Space Council, 24, 77, 78, 79, 81, 84, 86; Office of Management and Budget (OMB), 48, 49, 52, 69, 77; Office of Science and Technology Policy, 67; *see also* individual Presidents; President's Science Advisory Committee (PSAC)
Wilkening, Laurel, 67
Wood, Lowell, 81
Working Group on Extraterrestrial Resources (WGER), 55

Yeager, Chuck, 67
Yeltsin, Boris, 85
Young, John, 57

zero gravity—*see* weightlessness
Zubrin, Robert, 89, 90, 91
Zuckert, Eugene, 15

# NASA History Monographs

All monographs except #1 are available by sending a self-addressed 9 x 12" envelope for each monograph with appropriate postage for 15 ounces to the NASA History Office , Code ZH, Washington, DC 20546. A complete listing of all NASA History Series publications is a vailable at http://history.nasa.gov/series95.html on the World Wide Web. In addition, a number of monographs and other History Series publications are a vailable online from the same URL.

Launius, Roger D., and Aaron K. Gillette, compilers. Toward a Histo ry of the Space Shu ttle: An Annotated Bibliography. Monograph in Aerospace History, No. 1, 1992. Out of print.

Launius, Roger D., and J.D. Hunley, compilers. An Annotated Bibliography of the Apollo Program. Monograph in Aerospace History, No. 2, 1994.

Launius, Roger D. Apollo: A Retrospective Analysis. Monograph in Aerospace History, No. 3, 1994.

Hansen, James R. Enchanted Rendezvous: John C. Houbolt and the Genesis of the Lunar-Orbit Rendezvous Concept. Monograph in Aerospace History, No. 4, 1995.

Gorn, Michael H. Hugh L. Dryden's Career in Aviation and Space. Monograph in Aerospace History, No. 5, 1996.

Powers, Sheryll Goecke. Women in Flight Researc h at NASA Dryden Flight Researc h Center from 1946 to 1995. Monograph in Aerospace History, No. 6, 1997.

Portree, David S. F., and Robert C. Trevino. Walking to Olympus: An EVA Chronology. Monograph in Aerospace History, No. 7, 1997.

Logsdon, John M., moderator. Legislative Origins of the National Aeronautics and Space Act of 1958: Proceedings of an Oral History Workshop. Monograph in Aerospace History, No. 8, 1998.

Rumerman, Judy A., compiler. U.S. Human Spaceflight, A Record of Achievement 1961-1998. Monograph in Aerospace History, No. 9, 1998.

Portree, David S. F. NASA's Origins and the Dawn of the Space Age. Monograph in Aerospace History, No. 10, 1998.

Logsdon, John M. Together in Orbit: The Origins of International Cooperation in the Space Station. Monograph in Aerospace History, No. 11, 1998.

Phillips, W. Hewitt. Journey in Aeronautical Research: A Career at NASA Langley Research Center. Monograph in Aerospace History, No. 12, 1998.

Braslow, Albert L. A History of Suction-Type Laminar-Flow Control with Emphasis on Flight Research. Monograph in Aerospace History, No. 13, 1999.

Logsdon, John M., moderator. Managing the Moon Program: Lessons Learned Fom Apollo. Monograph in Aerospace History, No. 14, 1999.

Perminov, V. G. The Difficult Road to Mars: A Brief History of Mars Exploration in the Soviet Union. Monograph in Aerospace History, No. 15, 1999.

# NASA History Monographs

Tucker, Tom. Touchdown: The Development of Propulsion Controlled Aircraft at NASA Dryden. Monograph in Aerospace History, No. 16, 1999.

Maisel, Martin, Giulanetti, Demo J., and Dugan, Daniel C. The History of the XV-15 Tilt Rotor Research Aircraft: From Concept to Flight. Monograph in Aerospace History, No. 17, 2000 (NASA SP-2000-4517).

Jenkins, Dennis R. Hypersonics Before the Shuttle: A Concise History of the X-15 Research Airplane. Monograph in Aerospace History, No. 18, 2000 (NASA SP-2000-4518).

Chambers, Joseph R. Partners in Freedom: Contributions of the Langley Research Center to U.S. Military Aircraft of the 1990s. Monograph in Aerospace History, No. 19, 2000 (NASA SP-2000-4519).

Waltman, Gene L. Black Magic and Gremlins: Analog Flight Simulations at NASA's Flight Research Center. Monograph in Aerospace History, No. 20, 2000 (NASA SP-2000-4520).

www.ingramcontent.com/pod-product-compliance
Lightning Source LLC
Chambersburg PA
CBHW081451170526
45166CB00008B/2387